"The foundation of Cosworth was on the basis that it must be possible to make an interesting living from racing cars and engines," says Keith Duckworth, but it is a long way from Indianapolis to North London - four thousand miles, three decades, a pair of likely lads and a single ambition.

He first met Mike Costin in the mid 'fifties when the latter was one of Colin Chapman's bright young things at the emergent Lotus Cars in Hornsey. The student of mechanical engineering at Imperial College had purchased a Lotus 6 with Climax engine to replace an ancient Austin Ruby Tourer, the kit car meaning numerous trips to the factory for parts. This led to vacation work during university holidays, then the offer of full time employment as a Gearbox Development Engineer upon the successful completion of his studies.

The new B.Sc. soon struck up a rapport with the Technical Director, but all was not well at Lotus, as Mike explains: "He was working on what we called the 'Queerbox' - a gearbox with a very strange layout. Keith's analysis was that it would never work - which was perfectly right."

Despite drastic modifications, the five-speed transmission continued to baulk at genuine progress, so when matters came to a head, Keith decided to leave rather than attempt another stop-gap design around an existing stock of components: "I was not going to waste two years of my life developing something I knew would never work."

Duckworth had been on the Lotus payroll for only 10 months. During that time the professional respect for each other's talents had grown between him and Costin to the point where they contemplated the possibility of striking out alone.

Mr. Costin's reasons for wanting to leave Hornsey were more complex than a simple desire to be his own boss because they were strained, hectic times at Lotus; days of constant chaos and nights of lost sleep. "It was a stupid life, it was an illness. A couple of likely lads like us, somebody would want us to do something. It (the new business) would enable us to do our engineering, live out of it and get satisfaction out of life."

Initially, premises were shared with John Campbell Jones in West London and it was not until October of 1958, when the Company's very first real home was found at Friern Barnet, that Cosworth Engineering became formally incorporated. The business consisted of doing anything to earn some money, be it engine overhauls for the likes of Dennis Taylor with his Climax FPF, suspension modifications and even the bubble canopy for Vanwall as Mike's brother was its designer, Frank Costin.

Keith Duckworth at his desk.

As Cosworth grew so did the need for recruits such as Bill Pratt, still with them today as head of racing products planning. He describes Friern Barnet thus: "It was an old coaching stable adjacent to 'The Railway Tavern', in a total state of disrepair. An extension to it became the test shop. They still had the hayfeed baskets on the walls, the upstairs was totally unsafe and Keith used the bit under the stairs as an office. His missus came in at six o'clock, did him a meal, and caught up with the paperwork." These were the days of tuning the output of a small paraffin heater in an attempt to warm the farthest reaches, of chasing rats attracted by an adjacent seedy cafe, lunchtime swimming lessons at a local pool, and trips to Southgate tube station to use the toilet...

It was also the days of Formula Junior but plans to build their own Fiat-powered machine were shelved upon discovering the short stroke unit from the Ford Anglia 105E. A couple of examples were acquired and prepared, poste haste, for the 1959 Boxing Day meeting at Brands Hatch. The first Ford Cosworth, however, was not destined to end the day in glory, and there were other problems too. Keith explains:

"We nearly went broke in Christmas '59 because we could not make camshafts live in 105Es. We were down to having no money and no anything. Then just after Christmas I decided that the book on cam designs was a load of nonsense, so I produced a new design. It was the saving of the firm. That worked and it allowed us to win. Although we had disgraced ourselves at Brands, the first race of 1960, the Goodwood meeting, we were first with a Tyrrell Cooper second. Jim Clark drove the winning Lotus....That was the first cam to work reasonably well at high speeds and give a reasonable output.

"We dominated it from then onwards and put most of the other tuners out of serious FJ racing. We sold some to them at fairly high prices and they are still sold today as a sports touring cam for a 1600cc, the A2. The A1 was the disaster, the A2 was the one which worked. We went up to A8 over the next four years, gradually developing our FJ engines, winning most of the time then went into 1000cc Formula Two, which was the first cylinder head I designed."

Amongst those assisting with the 105E programme was Ben Rood who ran a machine shop at Walthamstow: "I first met Keith one day at Brands Hatch and thought that here is a bloke who has got some idea what it's all about. I started making pieces for the FJ and just got more and more involved."

Lotus had planned to make 25 Mk 18s and Costin's continued presence there helped Cosworth secure the engine contract for his alter ego. When demand exceeded expecta-

The early premises at Friern Barnet.

tions five fold, Cosworth were on their way. They obtained complete engines, discarded the unwanted items, then stripped, modified, built and tested the revised FJ specification powerplants. It was enough to save the Company, the only problem appearing to be the hordes of carbs, generators and manifolds which cluttered the premises as nobody had the heart to throw them away!

The vast improvement in business also had an effect upon Ben's activities. "I was working for nobody else. At one and the same go he (Duckworth) said 'How about joining us' and I said 'What about me joining you?' We just stood there looking at each other thinking 'that was a coincidence.' So I did, just like that."

Benny Rood became what he describes as a 'Self Employed Director', a situation lasting until after the eventual exodus from London whereupon he officially joined the Board of Cosworth Engineering. The former motorcycle and hydroplane racer/builder was now free to concentrate on his beloved engineering, not commerce.

Someone with no such distaste for paperwork was Bill Brown, a friend of Keith's from university. He took over the administration requirements of the now, fast expanding, company, and while his academic achievements at engineering were not great, his basic knowledge would be a great asset during negotiations until the mid-Seventies when he felt it was time to move on.

It was time to move on for the Company, now into the erstwhile Lotus premises at Edmonton where upgraded FJ and numerous special projects were soon supplemented by one of the most famous of them all, the Lotus Twin Cam, alias the Cosworth Mk XV. Installed in those glorious Lotus Cortinas - the forerunners of today's Sierra RS Cosworth in spirit and application - Grand Prix stars thrilled the crowds with their antics; cocking a wheel at the opposition, cocking a snoot at today's less adventurous superstars. Even now, the sight of an off-white Mk I with pale green sidestripes stirs the memories...

From the humble beginnings and dubious future, everything now seemed to turn to gold for the burgeoning company, so it was not too long before yet larger premises were needed. This time the decision was taken to find somewhere out of town where they could put down roots, settle and expand. They looked at a number of possibilities within their parameters before settling on a green field site at Northampton, in St. James Mill Road, known colloquially as Jimmy's End. Beyond the old farm gate, alongside the River Nene - less a babbling brook, more a turgid stream - the Cosworth empire would flourish and grow.

The initial buildings were approximately 7000 sq. ft with Bill Pratt and Brian Hart despatched in the summer of 1964 to oversee the setting up. Bill proudly explains: "We never stopped building or testing engines. We finished testing at Edmonton, then brought the next engine here for testing - a very smooth change." The staff level was then 14, but by the spring of 1965, the number had doubled.

Two new engines announced their arrival at Northampton, both 997cc, for Formula 2 and Formula 3 racing to be known as the MAE and SCA respectively. Formula 3 had been resurrected with the demise of Formula Junior with the MAE made in large quantities "like shelling peas" according to Ben Rood. If you stand very still at Hawthorns or Woodcote, you can probably still hear the shrill 'screamers' reverberating around the scenery.

Meanwhile the SCA, based around a Cortina block, was proving to be a very significant landmark indeed. Keith Duckworth explains: "I got into design from building because of being the only one around who could draw something, although Bill drew some early valves and things. I drew cams and then conrods, then cranks. It gradually

The present day premises at Northampton.

just built up so when the SCA came along it was the first cylinder head I had designed. That was a single overhead camshaft to follow on the best of our FJs. It won for two years before Honda with the four valve/wide angle head came and beat us. We had all kinds of combustion problems, it was too slow burning. I spent a lot of time thinking how you made this thing burn at a reasonable rate. I decided that it wasn't a Goer so we concluded the four valve head was the right approach. Having had a few false starts and tried it diametrically opposed - I gave it up at a fairly advanced set of schemes - I then drew the FVA which was conventional. I dropped the included valve angle to 40° whereas just about all the four valve engines had about 80° or 90° prior to that," before adding modestly, but with the sweeping rejoinder that is typical Duckworth: "I was the originator of the modern range of narrow valve angle/four valve pent roof engines. They had been made by Ernest Henry back in 1918 or so, but I didn't read about it. If you read you become ill-informed."

The way discussions about the SCA soon incorporate the FVA show the way the knowledge gleaned from one project is used to benefit the next and is typical of the forward march of Cosworth.

When the authorities announced a move to 1600cc Formula Two as from 1967, the Cosworth hierarchy soon realised there was ample scope to use a similar Cortina block in the new series as with the one it replaced. With the cylinder head derived from the SCA, with the notable addition of a twin overhead camshaft, this would be the FVA. At the same time Colin Chapman was looking to build a Grand Prix engine to contest the new three litre championship effective from 1966.

The Lotus chief approached his former employee with the suggestion of a V8, which suited Duckworth's design philosophy, especially with regards to weight and length as well as mechanical efficiency and fuel consumption. General consensus at the time favoured 12 or more cylinders, but as Bill Pratt remarks: "He is a great one for saying 'Forget what was good for my father or what has been done before. A new sheet of paper, I will start thinking.' He is a great thinking man."

One can only admire the faith shown in Keith Duckworth, his colleagues and their company despite the undoubted attributes and achievements. They had never before ventured into the heady arena of Grands Prix nor had they designed/built any engine from scratch. Always there had been someone else's block as a basic parameter.

Chapman tried to interest a number of parties including Aston Martin - a fascinating thought! - without success. Then he invited Walter Hayes, an old friend and then Director of Public Affairs at Ford, for dinner. It was a meal to savour.

Given the taste for the venture, Ford agreed to fund a 1600cc 16 valve/four cylinder Formula 2 unit based around the Cortina block, the aforementioned FVA, then a three litre 90° Grand Prix V8 using similar cylinder heads to its little sister. This would be designated DFV. The fact that Cosworth asked £100,000 for the two stage project is now motor sport legend, although less well known is the fact that Walter Hayes secured £150,000 from his board, having less faith in Cosworth's monetary sums than their engineering calculations. The extra funds were never called upon.

The initial priority was to build the FVA which Ben Rood describes: "Apart from the block, you could say it was totally our own design - and we even modified the block. The FVA marks the point where Cosworth engines took off, for the SCA was a triumph of development over design *(a favourite Cosworth saying this)*. It was the point where we made an engine and never looked back."

In the five years of top line competitive life before yet another rule change made it obsolete, the FVA was unbeaten while allowing great names like Jochen Rindt and Ronnie Peterson to hone their craft before moving on to greater things.

If the DFV remained competitive for a similar period, Ford would be happy because they realistically considered it feasible for only three or four years. Only time would tell.....

The FVA was the unbeatable Formula 2 engine, winning all Formula 2 Championships from 1967 to 1971 – the duration of the 1600cc Formula 2 regulations.

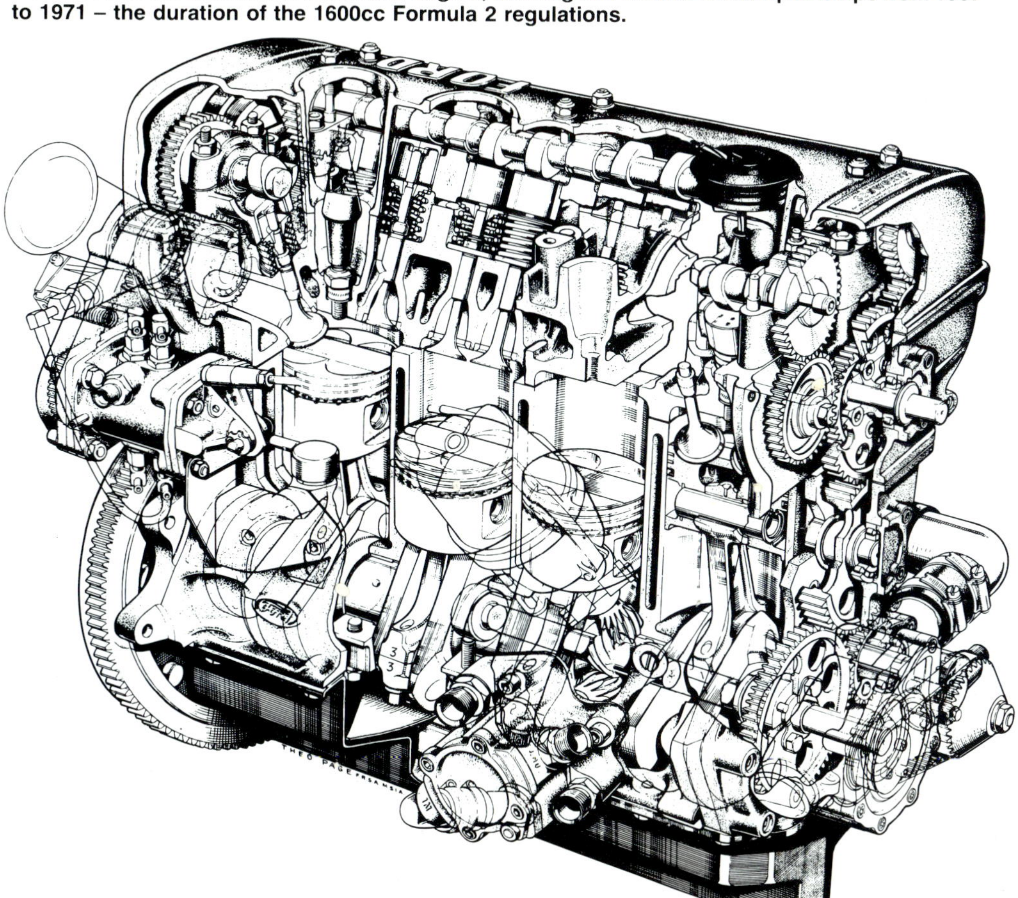

Writing the story of the DFV Grand Prix engine within the confines of this booklet is akin to summarising the history of the world on a postage stamp. Impossible! The story of the V8 *is* that of Formula One racing over the 20 years of the three litre series, so whereas tens of thousands of words cannot fully explain its true significance, the period can be accurately summed up in four: The Ford Cosworth Era.

When the car giant agreed to fund the programme, everyone seemed happy. Chapman had an engine to pursue his Formula One aspirations, Cosworth had money in the bank and Ford an entree into the glamorous world of Grands Prix. If all went well, Lotus and Clark could envisage untold glories, the small Northampton firm would race into the big time and the FoMoCo marketing men could finally put the lid on those epithets of Dagenham dustbins! All it needed was for Duckworth, Costin, Rood and co. to deliver the goods. Quite a responsibility for a team who had never, ever started with a blank sheet of paper before.

"I wasn't going to proceed in designing the DFV unless we got over 200bhp out of the 1600cc. We only needed about 400 from a Formula One engine to win, therefore doubling it.

"You take out your sheet of paper then realise how much more difficult it is than putting a head on it where you can start by drawing the bolt holes. You may not like where they are, but at least you can dirty your paper to that extent."

Gradually, though, ideas evolved in mind and sketch with the first commitment to the rear face and starter so ZF could commence work on the transaxle. Duckworth had always maintained that the cylinder head is the most significant single aspect of any successful engine, so many hours were spent refining theories before committing them to the drawing board and ultimately into metal. The FVA was already proving itself in competition and a number of 1500cc FVBs constructed for more direct comparison.

As the boss concentrated on matters at home, down at the factory former BRM penman Mike Hall detailed the ancilliaries aided by a couple of 'helpers' while the job of bolting together the first DFV would become the honour of George Duckett.

Eventually it was ready, a 2993cc whose 85.7mm bore and 64.8mm stroke was to become the standard for years to come. It weighed in at a full ten per cent less than target at about 360lbs while initial testing brought achievement of its first performance goal when the DFV reached 408bhp at 9000rpm, although not without difficulties with oil accumulating in the cylinder heads and susceptibility of the gearwheels.

More immediate was the clutch problem when installed into the Lotus 49 that Chapman had designed specifically to take the new engine. Mike Costin, who had the honour of doing those initial runs in the car as both Clark and Graham Hill were unavailable, takes up the story: "We took it up to Hethel just to thrash it up and down the runway as it was not a circuit in those days. The clutch flew apart, so we took a Cosworth twin plate out of a Team Lotus Cortina and it became the DFV clutch for the next five years."

While the bits of the original were returned to its manufacturer with a terse note, the programme went into a second day, this time at nearby Snetterton. Despite severe handling problems, Mike saw the potential of the combination. "It was impressive. Our old F2 car (a Brabham retained for development purposes), the fastest it would do was about 160/165mph. This thing was roaring up, doing about 175mph and still accelerating like stink with the sleepers at the end coming up a bit sharpish. At that sort of speed they start to look very hard about halfway down the straight!" It was the only occasion he ever drove the car, Hill taking over the limited testing available before debuting at Zandvoort for the Dutch GP. Team-mate Clark never even saw his steed prior to commencement of the meeting.

Suspension difficulties restricted the Scot to the third row of the starting grid, while Hill showed he had no such problems by snatching pole position from Gurney's Eagle-Weslake. Although blatantly unrealistic to propose a first time victory for the new venture, nevertheless realists contemplated and dreamers dreamt: Sunday would be make or break.

For Graham Hill it broke. The suspect timing gear failed early on so lowering the chances of success, but heightening the tension as Clark made up ground, took the lead and reeled off the laps. Cooly and methodically he went about his business until, it seemed like an eternity, the chequered flag came down on an historic victory. Clark had done it, Ford had done it, Lotus had done it, Cosworth had done it. Hurrah!

The famous victory at Zandvoort in 1967 where Jim Clark in the DFV-powered Lotus 49 clinched a first time out win for the engine.

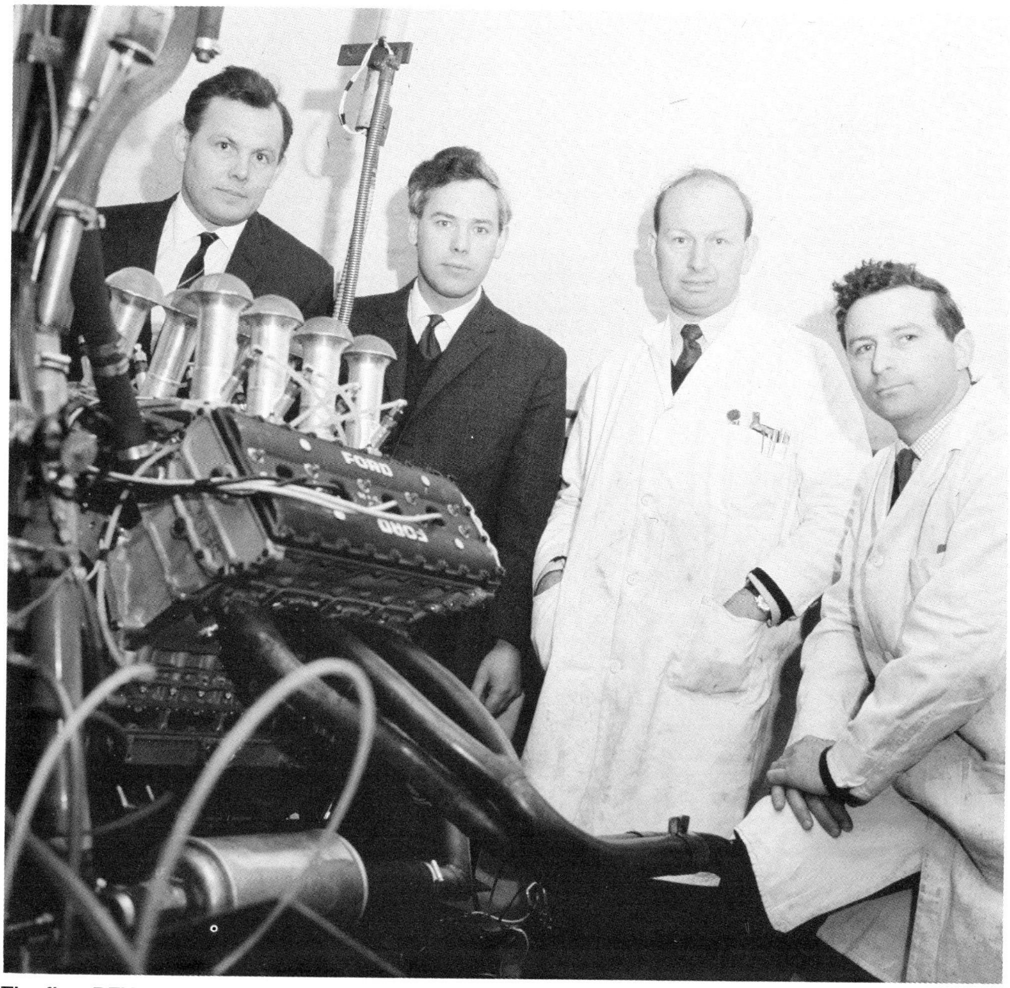

The first DFV on test with Bill Brown, Keith Duckworth, Ben Rood and Mike Costin looking on.

were having various failures despite continuous improvements."

The star pairing of Hill and Clark secured pole position and led each of the nine events they undertook that summer to register four victories, all to the mercurial Scot, but insufficient to stop Hulme and the Brabham-Repco taking both the Drivers' and Constructors' titles.

1968 started well with an excellent win from pole position for Jim Clark in the South African GP. This race, on New Year's Day, heralded not only the final manifestation of the traditionally liveried Lotus before Gold Leaf's coat of many colours signalled a whole new dimension to the sport, but also the last Grand Prix success for Clark. Before the combatants were re-assembled for the next race, at Jarama in May, the sport would suffer a great tragedy when he was inexplicably killed at an insignificant Formula 2 race.

If Kyalami was the end of some eras, it also marked the start of another as the DFV was now available for purchase by others and there was no shortage of those willing to pay £7500 for the privilege. Alongside Clark and Hill on the front row sat a similarly powered Matra, driven/managed by Jackie Stewart and Ken Tyrrell, who in the next few years would take over the Chapman/Clark mantle.

Lotus recovered from the loss of Clark in the best possible way with Hill winning in Spain, then at Monaco. By the time he notched up his third victory at Mexico in the final round, both championships were assured. The DFV was a World Champion in its first full season. Other wins by customers served to swamp the points standing for the

Graham Hill giving the DFV its first outing at the Lotus test track at Hethel prior to the 1967 Dutch Grand Prix.

The euphoria was understandable and also the relief when the subsequent engine stripdown discovered a missing gear tooth, the same problem which had sidelined Graham. The fates had been kind to Cosworth *et al.*

Although the Zandvoort win was the stuff of legends, it was not the case of riding off into the sunset and living happily ever after. Throughout the remainder of the season the Lotus 49 suffered a variety of suspension maladies, the matched ZF transmission ill-chosen and the DFV had its own problems. The designer recalls: "We had continuous trouble with the valve gear, the cam drive, until 1971 when I produced this weird and complicated cushion hub in the middle of the gear train - known as the second compound gear - which fitted in the original place. We could modify early engines. That got us over most of our valve gear troubles because otherwise the engine would have died as we

Ford Cosworth brigade, one lone Ferrari win being the only resistance to a clean sweep. Not at all bad! The good times rolled on into '69 where neither the Scuderia nor anyone else were allowed a consolation prize as the Double Four Valve thundered to victory in every round with Stewart's six wins culminating in a deserved first title for the new Scottish hero.

A few years before, the might of Ford had crushed Ferrari at Le Mans. Now they were doing the same thing in the Grand Prix arenas around the globe. Veni, vidi, vici.

A new roots-type scavenge pump effectively ended problems about blow - by oil trapped in the heads under pressure, while a spate of broken crankshafts was traced to a machining error. These and many other problems kept Northampton busy for this was still no easy ride despite the outwards signs of success. Much like a duckpond, it may appear calm on the surface, but beneath the waves the ducks were paddling like crazy as engineer Malcolm Tyrrell remembers: "Mike, Keith and us lads slumped in the corner of the test bed, at about 10pm, absolutely bemused and wondering what to do."

Whatever they did, they did it well, for the DFV continued to win.

Rindt won the championship in 1970, albeit posthumously, with the Ferraris at last mounting an effective response. The resurgence of Maranello, the piercing scream of the Matras - more sound than fury - and the last hurrahs of BRM's turbulent history all fought to overthrow the kingmaker, the DFV. The unholy trinity of 12 cylinder engines could win battles, but not wars, although throughout '71 it seemed that only the new Tyrrells stood firm against the onslaught. Such was Jackie's dominance that his supreme talent allied to an excellent first model from the Ockham team was enough to secure the title before the year's final races were run.

Tyrrell's transformation from consumer to constructor seemed to act as a flagship for new names being attracted into the sport, courtesy of the opportunities afforded by the V8. Whereas only a few years earlier the arena was reserved for a privileged minority, the arrival of the DFV and the 'knock-on' effect made it a virtual sport-for-all as the likes of March, Shadow and Ensign stepped into the glare of Formula One to wave the banner of Grand Prix racing. The arrival of new teams and new constructors also meant a proliferation of other talents, notably drivers and designers, who first made their mark accompanied by the pride of Northampton.

Emerson Fittipaldi's two crowns flanked a third for JYS. In 1973 the challenge to the ever-evolving DFV was virtually non-existant as the twelves went into decline again. The French had departed, BRM were bowing out ungraciously and Ferrari having one of those seasons where the Prancing Horse was

After Lotus had exclusive use of the DFV in 1967, it was made widely available and for the next 14 years it was to form the backbone of Formula One racing. One of the first teams to avail themselves of the engine was McLaren who in 1968 had recruited reigning World Champion Denny Hulme who scored his first win of the year at Monza (below).

The classic DFV x-section.

about as effective as a gelding at a stud farm....

It all proved to be the lull before the storm. Niki Lauda arrived at Maranello bringing about, almost single-handedly, the biggest threat the V8 teams had encountered so far. The ebb and flow of success was now firmly with the scarlet machines although it was partner Regazzoni who came closest to denying Fittipaldi his second championship. The Austrian made no mistake the following year and would probably have concluded a hat-trick except for that notorious Nürburgring accident of '76.

As the mid-seventies proved to be a golden period for Ferrari, sceptics seized the opportunity to pronounce the DFV's obituary - it had, after all, outlived the expected life span quite considerably - wreathed not in victory, but in memoriam. Some teams went so far as to introduce alternative power units: the Alfa Romeo F-12 was seen in various Brabhams of the genre, the Matra was resurrected without success and Renault introduced a 1½ litre turbocharged engine. The cynics cried that it would never win races....

While all this was going on, Cosworth implemented a major development programme for the apparent eclipse of the V8 caused great consternation at Northampton, Warley and Dearborn: winning had become a way of life. The result of these labours was a special edition of lightweights that became available for use by selected teams in 1977 with some benefits like low pressure crankshafts also incorporated into more standard engines.

Cosworth recall: "We actually made a complete magnesium engine for Tyrrell - mag block, sump, heads, carriers, the whole thing. The big problem was oil pressure because magnesium grows and distorts at a tremendous rate. The whole essence of a DFV has been getting tight bearing clearances. That thing grew so rapidly." Although the project was terminated, hybrids with aluminium blocks and sumps continued while Jody Scheckter used a conventional unit when scoring the DFV's 100th win at Monaco '77 - a decade all but a fortnight after the first.

The tenth birthday celebrations were spoilt by another Lauda victory, this time at Zolder, but now, after a couple of years in exile from the victory rostrum, innovator Chapman was bringing Lotus back to the forefront with another masterstroke: ground effects. The use of the underneath of the car as an aerodynamic force breathed new life into the DFV as the narrow base of the Cosworth creation was ideally suited to the medium whereas the wide twelves restricted the all important venturis. And it proved provident for while the classic pairing of Andretti and Peterson totally dominated the 1978 championship, next up in the series came Ferrari, then Brabham-Alfa Romeo - with and without the controversial fan device - then the other Ford runners. All things being equal, the Lotus was usually unbeatable in '78.

Unlike in 1979. Chapman attempted to stay one step ahead of the opposition by phase two of his theory, shaped like the radical Lotus 80, but it proved a costly and time-consuming error. Also, the others were quick learners, notably Ligier and Williams, against the Ferrari whose sheer power was able to make up for some of the aerodynamic disadvantages of its design. Ferrari driver Jody Scheckter took the title with his teammate Gilles Villeneuve as runner-up.

In order to take maximum advantage of ground effects, it was decided to drastically alter the auxiliary pumps at the base of the block along similar lines to those now operating on the DFX Indycar derivative. A single uprated water pump replacing the original pair and a smaller oil canister were the most notable changes. When the DFV had been conceived, ground effects applied to gardening not racing cars!

The modifications assisted the likes of Williams and Brabham in their battles against the ever more potent Renault, all the time the turbos becoming quicker, more reliable, more threatening. The DFV, so long the mainstay of Grand Prix racing, was being undermined by an arbitrary equivalency factor inserted into the regulations all those years ago. The immovable object was being pushed aside by the irresistible force. Soon Ferrari would join the turbo ranks, then BMW, then others. Keith Duckworth, who would eventually be required to join the turbos, remains steadfast in his opposition to them to this day: "The rules only allowed supercharged engines which is something that increases pressure on the inlet side. It does not allow you to increase the expansion side of an engine."

Years of anguish for Frank Williams finally paid dividends in 1980 as Alan Jones routed the Drivers' title and with the aid of Reutemann clinched the maker's version for good measure. The next year the melancholy Argentinian had an excellent opportunity to conquer the world himself, but gambled and lost in Las Vegas, Piquet coming up trumps.

In 1982, 15 years after its debut, the DFV was still winning regularly with seven different drivers taking the honours with Ford Cosworth power to back them up. Amongst them were Lauda, three races after his return from retirement, now adding to the DFV score instead of fighting against it, John Watson's epic drive in Ford's backyard at Detroit and Alboreto's finale in the Nevada parking lot where fifth place for Rosberg was enough to clinch his title and the 12th time the mighty DFV was the motive power for a World Champion. It was to be the last as the rapidly burgeoning 1500cc turbocharged engines finally took advantage of their small print benefits to obtain power outputs far beyond the realms of any normally aspirated unit.

Despite the onslaught from the turbocharged brigade, the 3.0 litre DFV was still able to win races and championships to the end. In 1982 Keke Rosberg became World Champion in the Cosworth-powered Williams FW08.

First time out, the Lotus 49 with the 1st DFV installed in the back.

Before the McLaren turned to a turbocharged unit they were successfully using the DFV in their chassis, as seen here in Niki Lauda's car in 1982.

Cosworth made one final effort to stave off their advance with a two fold development known as the DFY. The initial project notably resulted in a short stroke/large bore with modified camshaft, bigger ports and larger valves. It was really an advanced DFV while the 'pukka' DFY came later with narrow angle heads and integrated cam carrier/cylinder heads which helped drop the weight below 300lbs for the first time. They were rewarded with three wins in 1983 - albeit, all at street circuits - including the ubiquitous V8's final victory aback Alboreto's Tyrrell at Detroit, an appropriate finale for the Ford funded engine. The encore came two years later when another Tyrrell, this time driven by Stefan Bellof, scored the last ever points with a fine fourth place at the same venue. The Surrey team was, by then, Cosworth's last customer, but soon to join the ranks of the turbo brigade themselves. For 1986 the controlling body, FISA, decreed only turbocharged engines would be eligible so the reign of the DFV V8 had come to pass. Or so we thought....

It had scored a magnificent 155 wins over nearly two decades and a full dozen victorious marques. The widespread availability enabled many to compete who could otherwise never have envisaged contesting the ultimate series. The quality and competitiveness of the product created close matched racing for the benefit of accountants, combatants, sponsors and spectators alike. It gave the formula stability to build the foundations for the high profile, highly professional Grand Prix series that exists today. The Ford Cosworth era.

The 400th DFV was delivered to Heini Mader while the engine's 'godfathers', Keith Duckworth and Walter Hayes, look on.

The DFY (right) was a development of the DFV which scored 3 wins in 1983.

It is said that as one door closes another opens, and so it must have looked as Cosworth battled to stay ahead of Lauda's Ferrari and then the turbo cars in the Grands Prix of the later seventies.

The business was based on just a couple of racing engines, albeit it very successful ones, so with the possibility of the DFV being a fading flower in Formula One, it was imperative to look for new fields, new markets.

"Bill Brown recognised the potential," says Mike Costin. "We made six sets of pistons and camshafts before Vel's Parnelli Jones (the US racing team) came along to buy some crankshafts. So we flogged them to 'em. They did all the development work themselves, taking a standard DFV and converting it into a DFX. We were forced to move carefully and, in hindsight, it was the only way."

VPJ commenced development of a shortstroke/turbocharged unit in 1974 as a reaction to horrendous reliability problems with the Offenhauser that for so long had been *de rigeur* in USAC racing. The main engineer to the project was Larry Slutter who would later head Cosworth's own US engine shop, with help from crew chief Jim Chapman and Dick Scammell who arrived in California to work with the GP team having left Cosworth after helping with prototype assembly at Northampton. Dick was to stay in California for two years before returning home and to Cosworth and is now a director with special responsibilities for all racing engines, including the DFX.

It must have been purgatory for him to witness them pulling his beloved Cosworths apart - especially when pistons were filed to ascertain clearances! Unlike their Grand Prix counterparts, the Americans are used to servicing their engines with most teams having their own rebuild facilities. This was different, however, this was development.

To compete with USAC regulations, the capacity was reduced to 2645cc by limiting the stroke to 57.3mm then matched with a variety of single turbochargers. Standard Formula 1-type cylinder heads were initially used, although it soon became obvious that they could not cope with the prodigious outputs of power now generated with insufficient water circulation creating overheating and adding to problems with valves/seats until revised heads were designed with improved water flow and exhaust valves modified to put them on the same centres as inlets. Difficulties with local pistons would remain for quite some time. Magnetos were a problem and a spate of split hoses all hindered progress. The ancilliaries were substantially altered, notably a bigger scavenge pump which would eventually be transferred to the Grand Prix version. In some instances the teacher would learn a lot from the pupil.

Many of these modifications were still well into the future when the DFX first ran on the dynamometer, showing over 800bhp at 9500rpm with 80" Hg boost. So were many more after Cosworth eventually took over the project bringing their own pistons etc. to improve reliability from those tentative, early days.

In fact the number of difficulties turning a Grand Prix sprinter into an oval racer were quite small, a credit to the initial design, so it was with guarded optimism that one was installed into a Parnelli VPJ-6, more a skinny tyred F1 than a genuine USAC spec. machine, Al Unser doing a brief shakedown test before he and Mario Andretti practised for the 1975 Indianapolis 500, the big one. Unfortunately the car was far from raceworthy and subsequently withdrawn.

Duckworth comments: "If we had gone and won first time out I reckon it would have been banned. We had a boat that won first time out...."

With the demands of the local F5000 scene and a fraught Grand Prix schedule to upkeep, the team decided on a three race DFX programme that year, the second being Pocono when piston failure terminated practice while the DFX debut was finally made in the Bricklyn 150 at Phoenix. Bad understeer restricted Unser to ninth spot on the grid and an eventual fifth place at the end, one lap behind the legendary A.J. Foyt with his Coyote Ford V8. While the chassis remained difficult, the new engine was proving competitive and fuel efficient. All it seemed to need was reliability. Already other teams were showing considerable positive interest and the USAC powerbase was starting to move.

1976: fourth at Phoenix, seventh at Indy after rain stopped play and another fourth at Milwaukee. The reliability was now coming. Next on the agenda was the prestigious Schaefer 500 at Pocono in Pennsylvania. Heavy rain reduced the grid to being drawn by lots, Unser getting a short straw to start from the middle of the 33 car pack.

From the start he swiftly made progress, getting into the lead at 20 laps before being delayed by pitstop muddles prior to a rear tyre blowing at high speed thereby forfeiting nearly two laps. The situation looked unretrievable, the position hopeless. Far from it! Al surged back into the race, retaking the lead just after half distance with his glorious engine now attuned to a chassis modified by John Barnard. The performance made the might of the American racing scene look positively agricultural as Unser strove towards the chequered flag, a late misfire failing to deny him his destiny.

As America celebrated 200 years of independence, the British had won a major battle. The Americans would learn to revere a new name - Cosworth. The Cosworth era, mark two, had begun.

Al scored two more wins to clinch fourth place in the championship. Danny Ongais then joined the team for a two car assault in 1977 when the opposition was going to include McLaren M24s for the 'works' team, and Penske Racing. The word was beginning to spread. Five drivers won eight races that year helped by DFX power with Tom Sneva clinching Cosworth's first States-side title, but the big prize, Indianapolis, eluded them with Sneva and Unser trailing home behind that man Foyt. At least Ford were happy!

It finally came right in 1978. Five drivers again scored DFX wins, this time 14 of them as the pendulum began to swing right away from the local engines. It was only right that Al Unser should score the first Indy win for Cosworth, having been with the project from the start. During the course of the season he would become the first, and only, man to win all three classic 500-milers (Pocono, Indy and Ontario Raceway in California) in the same

Danny Ongais' Parnelli VPJ 6B-Cosworth as seen in 1978. A milestone was achieved with Al Unser's win in the 1987 Indy 500, 10 years after the engine's first win there (overleaf).

BOSCH
HOLSET
TURBO
Cumm

25T
Miller
AMERICAN
COMPUTERVISION
MONROE
ns
GOODYEAR
J. WILLIAMS CARS
25

year, although it was not enough to stop Sneva retaining his crown. Not that things looked good a few weeks before Indianapolis when Al destroyed the Jim Hall-prepared Chapparal Lola 500 in a massive accident. All was well on the day with DFXs filling four of the top six slots and leading every lap but one (during the pitstops) in a crushing display for all America to see: live, nationwide, coast-to-coast. For three years since those tentative steps at The Brickyard, the call had been 'The Cossies are coming, the Cossies are coming.' Now, for sure, they had arrived.

The DFX revolution on track was augmented by political struggles off it leading to the teams, CART, forming their own championship for 1979 to be run totally separately from the establishment's USAC series. Gradually over the course of the next few years, the traditionalists would relinquish control of all but the Indianapolis 500. Even today it remains a single USAC jewel in the CART crown.

The DFX became an innocent victim of the struggle with a myriad rule changes. This applied particularly to USAC events where they were attempting to apply a tourniquet rather than a humane coup-de-gras. The most significant change was reduced boost levels although, ironically, this only served to benefit the increasingly more refined DFX with added reliability.

The instigator responds: "We were knocked progressively down. It required us to be ingenious enough so that every time we found more power."

Then came the era of Rutherford with his Chapparal 2K against Mears and Bobby Unser in their Penskes, the trio scoring a massive 29 victories between them over the three years until the close of the '81 season. That year also saw a new name on the winner's rostrum - March Engineering with the Bicester company also getting fastest qualifying time for Indy via Tom Sneva, a portent of things to come.

As the eighties unfolded it was a case of not if a DFX would win but which one, the 1982 Indianapolis grid, for example, having just two intruders in the line up. The Cosworth just kept on winning, winning and winning. New names emerged as heroes, men like Bobby Rahal, while refugees from the Grand Prix circus also took to this dynamic form of racing, men like Derek Daly and Roberto Guerrero. Emerson Fittipaldi came back and Mario Andretti came home with his son, Michael, jousting with other famous offspring like Al Unser Jr. and Geoff Brabham for their turn in the spotlight. The DFX had a similar effect on motor sport in the USA as it did in Grands Prix, the ability to produce efficient engineering which results in close, competitive racing.

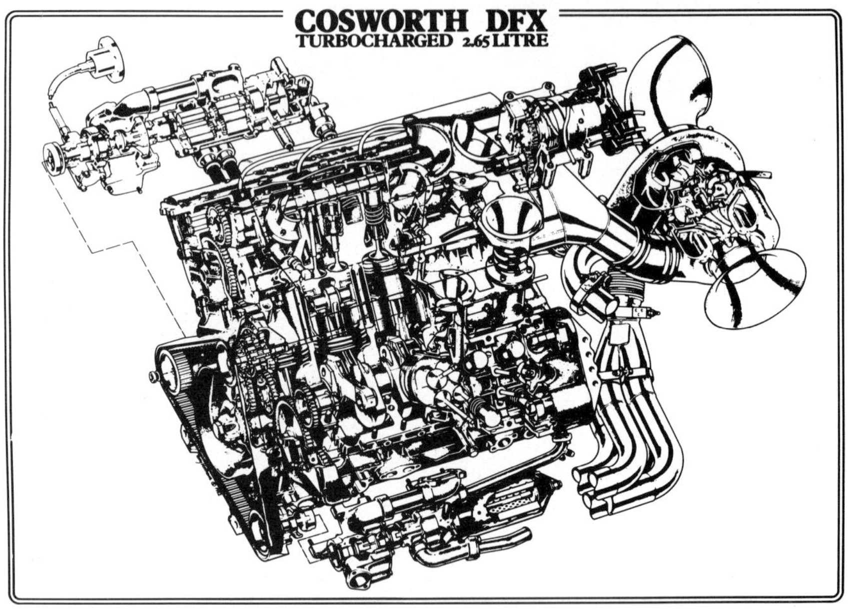

The 2.65 litre turbo-charged Cosworth DFX.

Nothing is forever. The first public awareness of a possible break in the status quo came during qualifying for the 1985 Indianapolis event when two Buick V6s headed the times to deny Cosworth their first pole position after eight years. The 3.4 litre Marches were aided not only by extra capacity but a USAC ruling of extra boost allowance for a 'stock block', thereby giving them about 800bhp, some ten per cent more than the DFX. The front row slots came as no surprise to insiders, nor did their inability to last more than 50 miles into the race: engine failure.

The Buick charge, not repeated in '86, seemed to act as a beacon for a whole host of contenders, all of whom were anxious to deprive Cosworth of their exalted position, their hard-earned position of supremacy. The threat from the East is underway with Honda already racing while there has been much written about the onslaught from Ferrari. Porsche, too, has announced a bid for glory far beyond the confines of Le Mans. Another already racing is the Ilmor V8, the brainchild of two former Cosworth personnel now engaged by Roger Penske and Chevrolet to bite the hand that fed them. Their unit, not unnaturally, bears a striking resemblance to its inspiration with all the advantages one would expect from a modern, purpose-built, contender.

The latest DFX produces some 700bhp at 12000rpm from a meagre 48"Hg boost, and will undoubtedly rise to the challenge of the Scuderia, Stuttgart or anyone else. After all, the others have a long way to go to emulate that fine record of nearly 150 wins, those pole positions, the fastest laps. Until then, when Johnny comes marching home, the chances are he will be driving a Cosworth.

Any engine that wins over 150 Grands Prix ranks amongst the finest technological achievements of all times. Any engine that wins that many 'Indycar' races can be similarly honoured. If the two happen to be one....Then, what if the same basic unit has also won umpteen sportscar events - including a brace of victories at Le Mans - and some saloon races too? What if successes were not limited to *terra firma*?

The achievements of the V8 away from Grands Prix and USAC/CART are enough to classify the DFV as a great engine for it has been many things to many men.

The first derivative was the DFW which Hill and Clark used to great effect in their Tasman Series campaign of 1967/68 aboard Lotus 49ers. The three litre displacement was decreased to 2491cc by the relatively simple matter of reducing the piston stroke to 54.0mm with shortened con rods. At the end of the 'Down Under' sojourn, they were converted back to DFVs to fight in front of a larger audience.

There are not many audiences bigger than those at Le Mans. The claim to be the first sportscar to install a DFV as its powerhouse is disputed, but certainly the first to participate at their spiritual home, in the *Vingt Quartre Heures du Mans*, was the 1971 Ligier JS3 Group 6 Spyder of constructor Guy Ligier and Patrick Depailler. Fitted with milder camshafts at Northampton and limited to 8800rpm the car carried French hopes in the era of Porsche versus Ferrari. Alas it was to no avail. When running well in the top 10 with six hours to go - despite a 90 minute pit stop to repair broken rear suspension mountings - it succumbed to gearbox failure. A 3½ hour rebuild saw it out on the track to take the chequered flag, although by now too far behind to be classified as a finisher under A.C.O. regulations.

The DFV notched up its first Sarthe finish 12 months later when the Brabham-based Duckhams Special of Alain de Cadenet and Chris Craft took 12th place. At lunch time on Sunday, looking poised for a superb fourth, bodywork problems followed by a damaging spin meant an hour lost, an opportunity forsaken.

1972 was notable for the late withdrawal of the factory Ferraris, the Scuderia considering their cars unsuitable for the rigours of 24 hours. Another to draw the same conclusion was the recently formed Gulf Research team headed by that sportscar immortal, John Wyer. It would be a different matter in 1975 when the GR8s of Ickx/Bell and Schuppan/Jaussaud came home first and third respectively to sandwich the similarly powered Ligier of Lafosse/Chasseuil.

Northampton had scored its first success at Le Mans with a resounding Cosworth 1-2-3. The A.C.O.-inspired fuel economy regulations - a forerunner to current Group C in many ways - had seen the DFVs reduced to less than 9000rpm for the purposes of economy and reliability, although at other venues the sportscar version of the Grand Prix engine was now pulling over a thousand more. A prudent move.

In 1975 the Cosworth-powered Mirages of Ickx/Bell and Schuppan/Jaussaud came a well earned first and third respectively at Le Mans. Another Cosworth-powered car came second.

The ADA Gebhardt powered by the 3.3 DFL was the C2 class winner in the 1986 Le Mans race. The DFL was also to power the Spice to C2 victory in 1987.

1987 saw the re-introduction of normally aspirated engines into Grand Prix racing competing for the Clark and Chapman trophies. Ivan Capelli, winner of the 1986 Formula 3000 Championship in his March-Cosworth, is seen in his March DFZ at the 1987 British Grand Prix.

Despite three cars in the top five the following year and runner-up spot in '81, the only other time the DFV managed to overcome the increasing stranglehold that Porsche was now exerting on the classic race was in 1980 when, as in any good Boy's Own story, local constructor Jean Rondeau became the first man to win the 24 Hours in a car bearing his own name. Partnered by Jaussaud, with a dash of '75 *deja vu*, another team car also took third place with Alain de Cadenet, always thereabouts but never there, in seventh.

The announcement that from 1982 sports car racing was to take on a new form was greeted with enthusiasm at Cosworth. The new Group C rules were governed not by cubic capacity but fuel allowance, so allowing scope for the company's inventiveness and the DFV's versatility to be given full rein. Although they had assisted the V8 sport car package for many years Cosworth were not, in truth, fully enamoured with making the sprinter run a marathon. Now they could develop a full endurance engine.

The programme was funded by Ford in much the same way the DFV had been born back in the mid-sixties, with the resulting engine known as the DFL. It consisted of a large bore 3.3 litre and a long stroke 3.6 then, if all was well, a large bore/stroke version of just under four litres. The project, in fact, virtually went from the smallest DFL to the 3.9 and it was there that the V8 encountered more problems than at any stage of its career.

Installed in the early Ford C100s and the latest Rondeaus, the 1982 Le Mans was a calamity for the DFL competitors. The 3.9 had a crankshaft damper which failed due to vibrations accentuated by the unique requirements of the three mile high speed Mulsanne Straight plus poor installation within the chassis.

Keith Duckworth: "It is well known that the V8 has a flat plane crankshaft, so fairly bad cross shake is to be expected. Taking it up to 3.9, although the stroke was only increased by half-an-inch, it made the vibration 45% worse. It did shake apart some of the cars where it was not well mounted, but the bad name as a shaker was due to the inadequacy of the mounting."

He is adamant about the potential of the large unit before the project was shelved by a directive from Ford: "I do not think it was ideal in its 3.9 litre form, but as a Group C engine it would have been very good. It wasn't big enough for normally aspirated (*an abundance of unused fuel proved that*) so it needed a turbocharger. It would have been a winner if the support and money had been around. It was fairly disappointing when it stopped."

The reference to the turbo recalls the experiment to produce blown engines that were ultimately to find their way into the 'works' C100s now under the auspices of Gordon Spice at nearby Silverstone. The model had marked the return of Ford to the arena where they had so much success with the GT40s, the J-cars and Mk. IIIs in the late sixties, although now their target would be German, not Italian. The original design had failed to cover itself in glory, but the revised version, by Tony Southgate, showed much more promise. The turbocharged engine actually undertook tests at the circuit before a policy amendment by the motor manufacturer pulled the plug and the C100 sank into oblivion.

When left beached by the decision, both Spice and Southgate went their separate ways in Group C racing with both collating a considerable amount of success. If only....

The 3.9 DFL seemed to have a jinx. As Group C got underway, the Rondeau won the first ever race for the category, at Monza, before the Porsche challenge was mounted. Then another had been the only Group C car to finish at the daunting Nürburgring when trailing home a Group 6 Lancia *barchetta.* The Le Mans fiasco was a turning point in their season, but the team appeared to have done enough to clinch the Makes' title for themselves and the Ford Cosworth by way of consolation. Unfortunately a little item tucked away in the small print gave 15 points to a lowly Group B Porsche 911T and the series went to Stuttgart.

In 1983 the authorities introduced a second category to Group C incorporating lighter cars with greater fuel efficiency for which the 3.3 has proved ideal. Now producing around 550bhp it has won the class at Le Mans with almost consumate ease through the likes of Rondeau, Gordon Spice and ADA Gebhardt. Much of the opposition, like the Giannini/Carma and the Austin Rover Group V6 has flowered, flourished and died. Group C Junior/C2 has seen some excellent racing reminiscent in many ways of those heady days of Grand Prix battles a decade and more ago, thanks to Cosworth V8.

These days the DFL interest in the top category seems confined to Tim Lee Davey's

The Ford C100 was powered by the 3.9 DFL but the project was stopped before it had accrued any success.

The 1987 Swiftair Ecosse was powered by the 3.3 DFL and was one of the more successful C2 cars winning their classes at both Silverstone and Brands Hatch. The Supervan (below) was powered by the 3.9 DFL.

privately funded and developed 3.3 DFL with turbocharger, as if trying to recreate the dream of the C100. After initial testing problems the engine, with its twin distributors, perhaps looks set to create a new chapter in the annals of Cosworth-powered history.

The demise of the 3.9 'shaker' was not the final hurrah for it either as an example was fitted into the Transit Supervan paraded by Ford as a publicity machine in 1984. The kevlar/carbonfibre bodied clone was capable of 165mph with an excellent power-to-weight ratio enhanced considerably by the fact that it was totally incapable of carrying any payload, the whole of the rear being filled with 'Cossie' and a five-speed transaxle! Following on from an earlier GT40 engined device, Supervan II increased its sports car associations by revealing that it was designed by Tony Southgate and the basis of the chassis was, you've guessed it, a C100.

If the Supervan seemed like an ideal way to use up an unloved 3.9 engine and an unwanted chassis, then other people have sought to use Northampton's finest in equally bizarre ways. A number of saloon car racers have seen DFVs shoehorned into them, notably a Capri and an Escort with Colin Hawker generally at the forefront of such exploits, while some intrepid soul managed to fit the V8 into a Volkswagen 411, to be known as the DFVW. Then there is the story that abounds at the factory of the 'serious' enquiry as to how one would install a DFX, all 700bhp and 400lbs of it, into a dragbike.

If a drag queen seems like a nightmare, then a powerboat must seem like a wet dream!

The most unusual installation of a Cosworth V8 must be the powerboat. Bill Brown was a keen powerboat devotee and through him Cosworth, in the person of Malcolm Tyrrell, got to meet Jackie Wilson who was keen to install a DFV into his craft and contest the 'OZ' Unlimited/Formula Libre category. Malcolm, who was acting as project engineer, advised against the V8 on grounds of cost, recommending instead the GAA V6, but was over-ruled by Keith Duckworth. The Company elected to help Wilson with a view that there was a whole untapped market for the DFV amongst all those boat people.

Results were encouraging with wins at places like Fairford and Chase Water, but they also got that sinking feeling when the boat went down at Amsterdam, Wilson clipping another contender at the chaotic start. With help from the attendant Dick Scammell, they salvaged it without undue damage and even managed to get it restarted, much to the surprise of all around - except the folks from Cosworth, of course.

If going Dutch was the nadir of the project, the zenith should have been the Parker Dam race, a seven hour endurance event, where it ran as part of the Mercury team in 1977. The vessel was, by definition, an outboard but when they got to Arizona the organisers deemed the DFV an inboard under local rules, thereby forcing the Cosworth three litre to be pitted against massive seven litre Chevrolets and the like. The Americans dismissed the puny little British attempt, at least until the race started, showing its heritage and coming home seventh overall behind the outboards of Molinari, Seebold and co. and first in class! They were 80 miles ahead of the nearest inboard rival despite question marks over the gearcase. The hosts were not pleased.

So much so that when it came to the prize giving, there were no accolades for Cosworth, the organisers deciding to disqualify the DFV on the basis that it was not an American built engine. As once was said about 'The America's Cup', Britain rules the waves but America waives the rules.

The Cosworth team were not amused, decidedly not amused, and Duckworth let rip with one of his legendary diatribes but to no avail so they *really* must have meant it! Nowadays he can laugh at the experience and he dismisses it with: "It was democracy. One man, one vote, and we only had one vote!"

The project for the next year, fired by the treatment at the Colorado River, was even more ambitious: to put two DFXs into a stretched Seebold catamaran against the likes of Circus Circus and Miss Budweiser in the Unlimited Hydroplane class. They were reckoned to be more manoeuvrable and have good power-to-weight characteristics compared to the American dreams.

The drawback of the DFXs was that their basic characteristics of good power high up but little low down was contrary to the requirements of getting a boat up out of the water. Mr. Duckworth suggested nitrous oxide. The power response was phenomenal but by then the steam had gone out of the programme as it was pointed out that the commercial future was non-existent.

The twin DFX machine was subsequently modified to take alternative units and passed on down the line. The DFV boat is now on display at the Wat Tyler Country Park in Basildon, Essex.

Another display item is the famous four-wheel-drive Grand Prix car the Company built in 1969.

Designed by Robin Herd shortly before he became a prime instigator in March Engineering, which would become a good long-term customer of Cosworth, the angular and unpainted device was certainly radical in a period of experimentation and false hope, destined never to race. Tested by Trevor Taylor and Mike Costin, the Cosworth director recalls those days without the enthusiasm evident when talking about the Lotus 49: "A good experience...five laps and you'd feel as if you had been doing press-ups all day...a good museum piece!" Precisely! At Tom Wheatcroft's splendid Donington Collection in Derbyshire.

Formula 3000 replaced Formula 2 as the stepping stone to Grand Prix racing when the two litre category was phased out at the end of 1984.

The DFZ as seen in Pascal Fabre's AGS at the 1987 British Grand Prix.

The incoming class not only gave the Formula One teams a very handy outlet for their unwanted DFVs, but also an upturn in the market for new examples. Limited to 9000rpm by the rules in an attempt to hold down costs, the formula was Ford Cosworth dominated right from the start, only a pair of Ralt Hondas presenting a challenge. Champions Danner and Capelli both proved the system worked by progressing into the top division.

When everybody thought the V8's story as a Grand Prix unit was over, FISA surprised them. It was announced that whereas 1986 was a turbo only season, '87 would begin the re-introduction of normally aspirated engines, with a curbing of turbo boost over the next two years culminating in their demise. 1989 would be normally aspirated only. The biter was bitten!

The news stated that the engines were to be up to 3½ litres, and Cosworth responded to enquiries by producing a 3494cc version by dint of the 90.0mm bore of the DFL and a stroke of 68.3mm, rated at 560bhp for 10250rpm. Amongst those seeking the revised unit were their old friends from Tyrrell, ready to win the Clark and Chapman trophies for their class.

The upturn in fortunes of the V8 as a commercial item, unseen at the start of the eighties, has had a magic effect on the sales. In 1987 Cosworth will sell more of the DF derivatives than in any previous year! Quite a coup for a 20 year old engine.

The places it has been and the things it has done make the Ford Cosworth V8 quite unique: Monza, Pocono, Indianapolis, Le Mans, Monaco; 24 laps or 24 hours, Grand Prix or Index of Efficiency; and messing about in boats.

Has anyone out there heard a rumour about a Cossie aeroplane?

The DFZ formula 1 engine (right).

Philippe Streiff in the DFZ-powered Tyrrell DG/016 which won for the team the first Colin Chapman trophy in 1987.

Although the V8 has received the major acclaim since 1967, there have been plenty of other goings-on deep in the heart of England; many a varied project, most very successful. Only most because, after all, nobody is perfect.

The DFV tends to overshadow the FVA which spawned it, but the 16 valve/four pot is a worthy champion in its own right. The basic criteria of the unit has already been explained, so suffice it to say that a 116E Cortina block was the heart and a Cosworth gear-driven DOHC became the soul. The Four Valve 'A' had the same bore as the DFV, 85.7mm, with a longer stroke of 69.1mm to make up the 1598cc. Initially it showed just over 200bhp at 9000rpm to give that vital impetus to the Formula One project, finally to top 230bhp at 10000rpm before being phased out by the rule book at the end of 1971.

Between times, the FVA dominated Formula Two in a similar way to its big sister in the Grand Prix category and a 1.8 litre version, known as the FVC, was doing equally well in the European Sportscar championship for up to two litres. Led by Jo Bonnier's Lola, the FVC pulled off two Euro titles in the four years of the championship. Meanwhile, the FVA crushed all opposition in the Formula 2 category, only a brief challenge from Ferrari with their ex-Formula 1 V6 'Dino' in 1968 and a handful of wins by BMW two years later preventing the Cosworth company totally sweeping the board for the period. The five year plan had been a good one. Ironically for the finale of '71 the biggest threat, in fact, came from their own BDA-type!

The BDA was Mike Hall's baby, Keith Duckworth being so involved with Formula One at that time. Mike explains: "We had chats about it, but things were left for me to get designed, developed and detailed."

Described as a 'productionised FVA' the BDA, Belt Driven 'A', was on a brief to use as much of the 1600cc Ford Kent engine as possible therefore all the bottom end, excluding pistons, was untouched. The top end was a different matter with the belt-driven overhead camshafts proving the ideal basis for an exciting road project to allow homologation into the Group 2 racing and rally circles of the time.

Hence the RS1600 Escort was born. The engine man continues: "It was a very high performance car, even by today's standards, because the BHP per litre of the BDA is not far short of what we are doing now with later designs. Two twin choke Webers, there were no constraints upon inlet manifolds as we have now with some fuel injected engines."

Rated at over 120bhp the RS1600 was about twice as powerful as most in its supposed category and started the whole trend of superfast saloons which remains until this day. Not only were Cosworth changing the face of competition, they were also moulding the future of road machines.

The BD-series, though, was intended for competition and soon the sporting versions appeared like the BDB for rallying and the BDC for Group 2 saloon car racing. In 1971 the new Formula Atlantic, a sort of Formula 2½, followed with its D-series while next in the alphabet came 1790cc and two litre types for the real Formula 2.

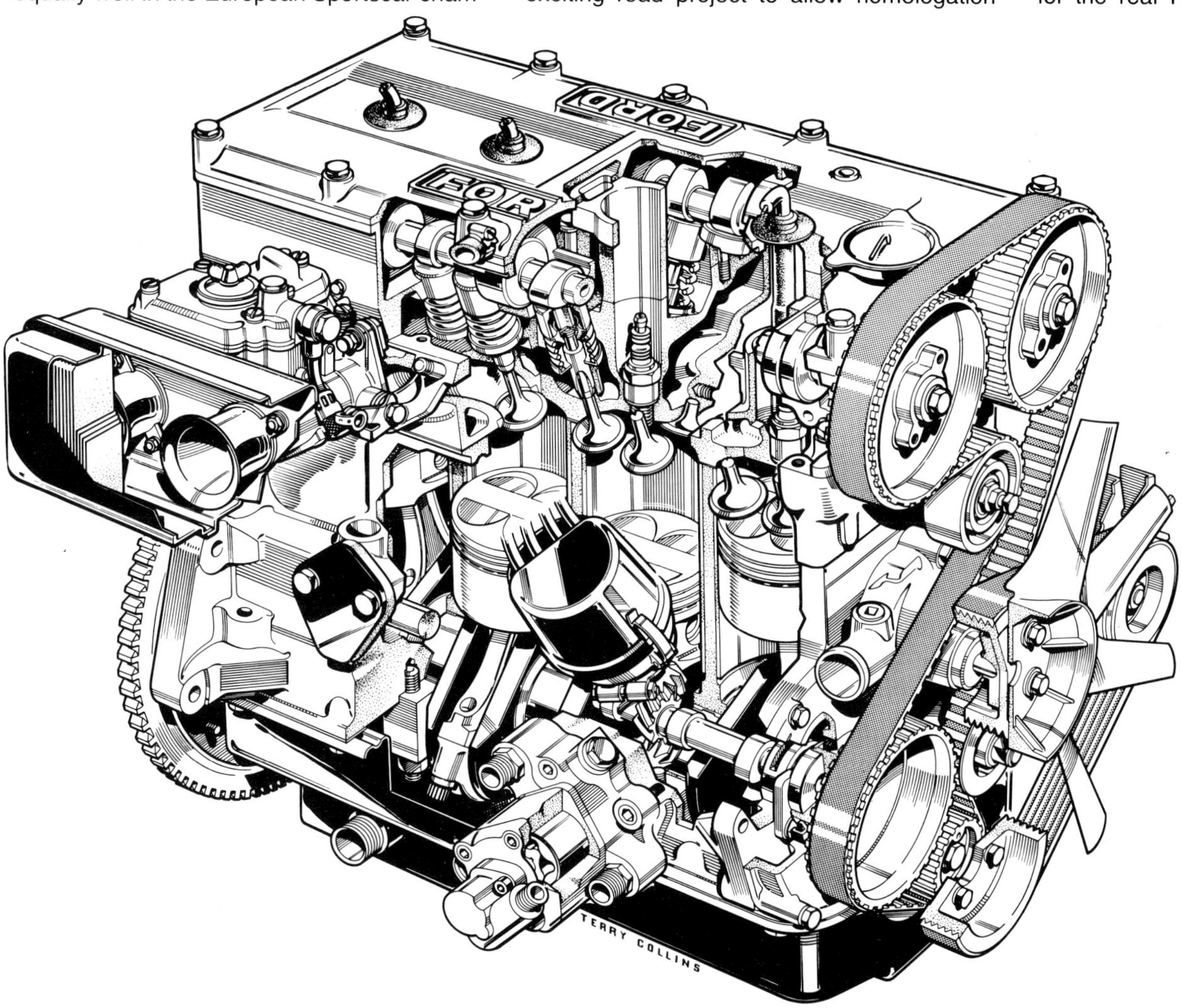

The BDA was a road-going engine designed for Ford for installation in the RS1600 Escort. Using the 1600cc cross-flow Ford block, its design was based on the FVA but with toothed belt drive to the twin overhead camshafts.

The RS1600, albeit with a two litre aboard, won three consecutive RAC Rallies for Ford, starting with the fine victory of Roger Clark and Tony Mason in 1972. Ford RS Escorts were to win the event for an unprecedented nine consecutive occasions, only denied double figures by Toivonen and his little ray of sunshine in 1980. Many other international rallies also fell to Escorts with BDs under the bonnet, notably Finland's very prestigious 1000 Lakes and the daunting Safari of both 1972 and '77. Neither the frozen wastes of the Arctic nor the shadows of Mount Kilimanjaro could thwart the diminutive car whose major rôle in life was to transport Auntie Ethel to the local supermarket. Today the international off-road scene may have moved on to more exotic machinery, but the club competitor still shows a healthy regard for the RS Escorts with their Cosworth-inspired motivation.

The RS1600 won three consecutive RAC rallies for Ford starting with the fine victory of Roger Clark and Tony Mason in 1972.

The 3.4 litre GAA-powered Ford Capris used in the 1974 European Touring Car Championship were good for 440bhp.

The machinery for the eighties was originally supposed to include the Escort RS1700T with its BDT turbocharged engine which was, in fact, 1800cc. Unfortunately the pace of development of rallying meant that after nearly three years work, it was obvious to the Ford hierarchy that two-wheel-drive was history if you were intending to win. The project was scrubbed, being replaced by the futuristic RS200 which carried the same engine, this time amidships, with 250bhp for road trim or over 400 in full Group B specification. Alas, this venture was similarly doomed when the controlling body banned all supercars after a series of accidents, Ford being left to dispose of a large number of vehicles.

One off track vehicle which did see competition in the eighties was Martin Schanche's 1860cc turbocharged rallycross device developing a massive 500bhp as it battled its way through tarmac, gravel or mud *en route* to dozens of wins. Nowadays there are replica machines, notably John Welch's example which can accelerate from 0-60mph in an astounding 2½ seconds.

On track exploits started with the European Touring Car Championship where well prepared Escorts won regularly in the 'up to two litre' division in the hands of exponents like Matthews, Brodie and newcomer Andy Rouse. These successes permeated down to national level in Britain and elsewhere with the Broadspeed and Zakspeed examples probably best remembered.

Single seaters were represented by the Atlantics, the UK's equivalent to America's F.B., and Formula Two. Vern Schuppan won the first F.A. title while others who benefitted from BDD experience included such top names as Tony Brise, Gunnar Nilsson and Alan Jones. The first season the BDA was properly eligible in Formula 2 was 1972 when Mike Hailwood clinched the Euro title.

Early engines had suffered reliability problems as tuners attempted to stretch what was, basically, a 1600cc to the full two litres, but it soon became evident that about 1840cc was the limit for the cast iron block. Later brazed cylinder liners, then an aluminium block, would overcome most of these problems although by now the massive resources of the BMW organisation were being mustered to dominate the category, thereby laying the ghost for all those FVA inspired defeats. Munich had learned well for their challenger bore more than a passing resemblance to its foe.

Other BD areas of operation have included the BDP for US midget racing with 230bhp from two litres by way of fuel injected methanol. This often overlooked aspect of American motor sport attracts loyal devotees to whom Cosworth have already sold over 80 engines, winning the West Coast championship in '86. Developed by the staff of the Company's Californian subsidiary, the BDP is continuing the theme that Mike Hall's design really is the motor sport engine for the masses having proved itself in a variety of arenas over the past 20 years. As the man says: "We built up a meccano set for the BDA with everything from about 1100cc to two litres. People could come along and buy various bits. They still do."

That Mike referred to the EAA, the Chevrolet Vega, as 'the stumbling block' was perfect, for the phrase amply sums up the debacle. "It was extremely light compared to the BDA with its pressure die-cast aluminium block from Reynolds, but was not up to the ratings we were putting through it. Designed to be a two litre racer, we could not make the cylinder block live. There was nothing we could do. It started as a race engine with Chevrolet, dry sump, the whole works, and it was Chevy who productionised it for the Vega car. They even had problems with the engine in road trim just as they did with the standard engine. The project died but they are now collectors' items!"

Thus the first road car to carry the Cosworth name did not enhance any reputations. It was not the engine which was at fault, only the block, but that tends to be somewhat critical and the whole episode fated before it began. Hundreds of blocks were being delivered, many failing a pressure test Cosworth devised for them and littering up the place while a solution was found. It never was.

Those which passed the test were released, the first version stirring in anger in Guy Edwards' Lola but failing to go the distance despite the attentions of Messrs. Duckworth and Scammell at Salzburgring.

It was the start of an ominous tale which resulted in the EAA never reaching sufficient status to go into Formula 2, Tommy Reid's hillclimb exploits heralding the greatest successes for this almost forgotten scheme. Forgotten that is until an example was recently found in one of Northampton's darkest corners. Perhaps it should be put on display with the epitaph that you cannot win all the time.

Alongside it there could be the JAA/JAB 750cc motorcycle twin cylinder developed for Norton Villiers in the mid seventies and described by Cosworth as a 'sad and sorry saga.' The least said about that the better.

Looking down the list of Cosworth engines, the memories are again stirred by another of Mike Hall's projects, the GAA. Based around a V6 'Essex' cylinder block, there were about 100 of the 3412cc, twin ohc/24 valvers made, belt driven. These became the basis of those mighty Cologne Capris that did so much to enhance the Group 2 saloon races of the early seventies when the likes of Jochen Mass and Klaus Ludwig fought the BMWs for supremacy. 440bhp at under 9000rpm made the GAA a formidable beast and many were later installed in single seaters to take on the brutal Chevy V8s in Formula 5000, their relative compactness and lightness making up for any deficiency in outright power.

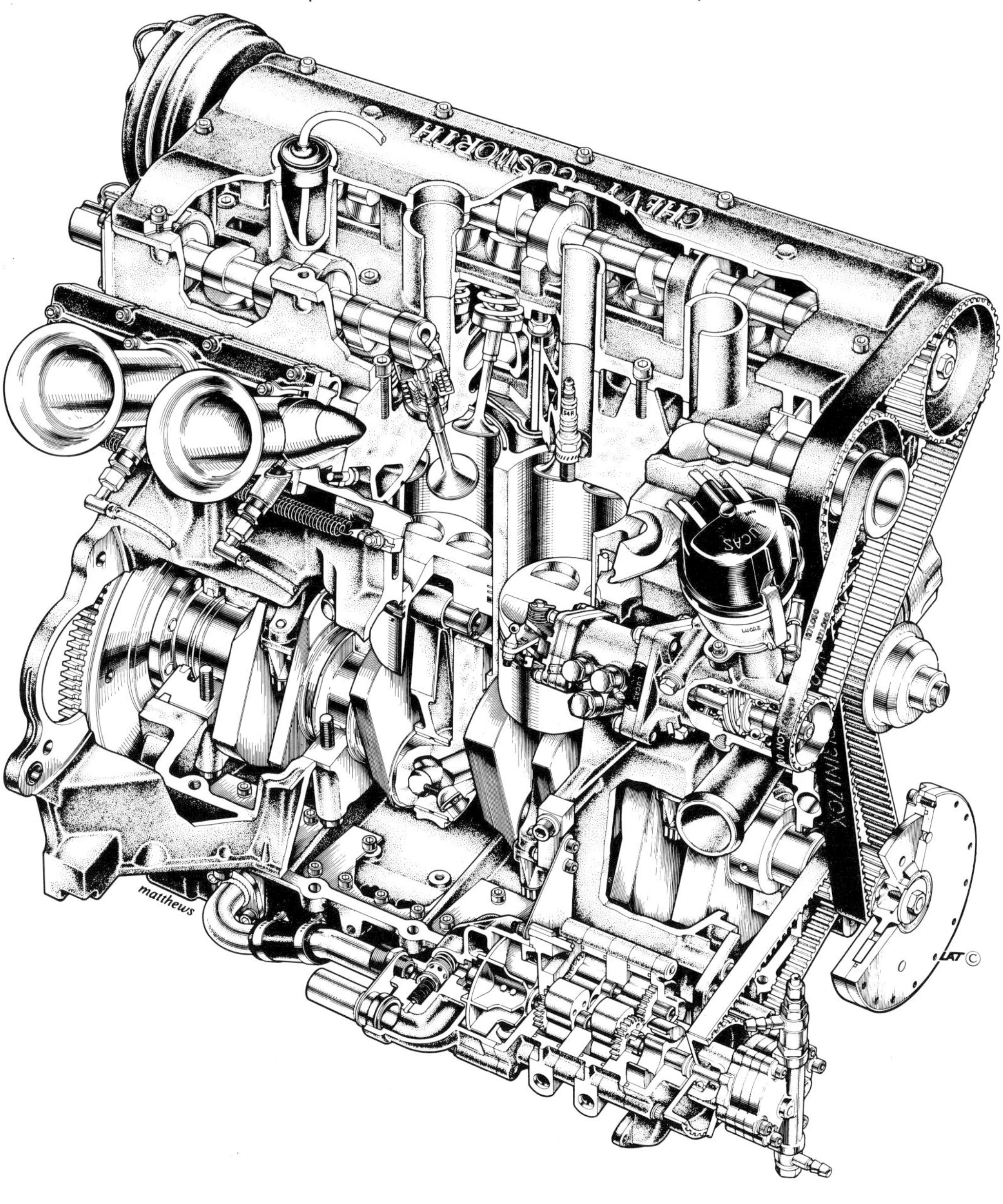

The 1995cc EAA engine was based on the Chevrolet Vega block and was Cosworth's first non-Ford based stock block engine.

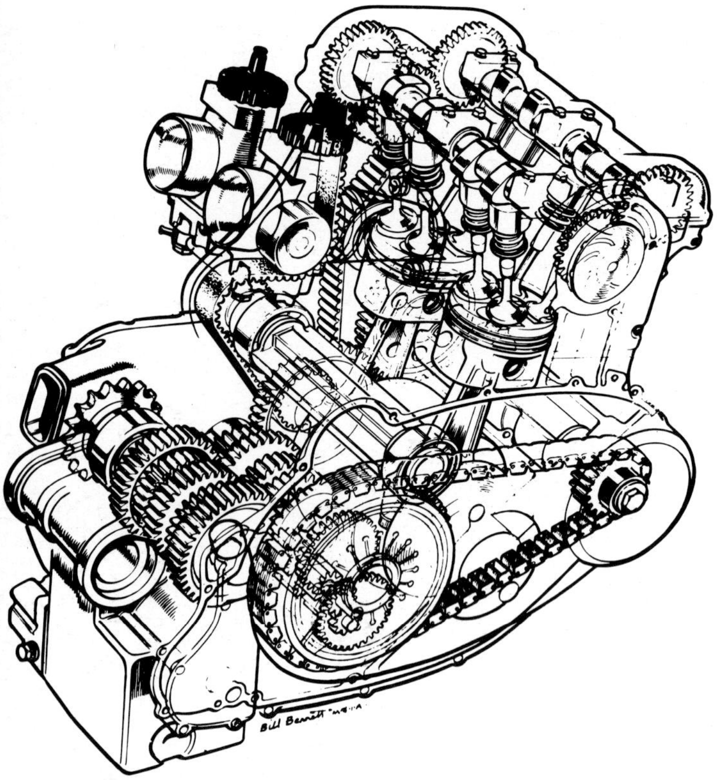

The 750cc JAA was a production motor cycle engine designed and built for Norton Villiers.

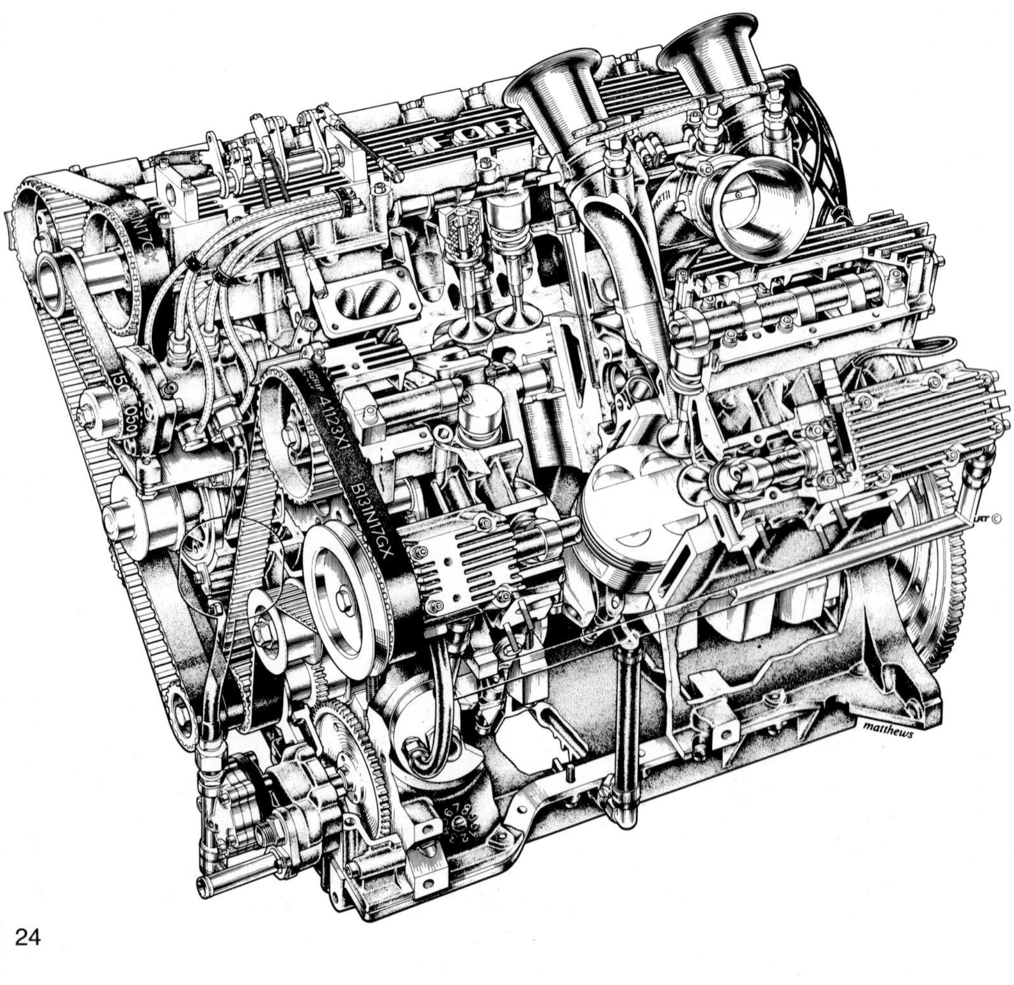

The GAA was a Ford V6 block based engine for use in Group 2 saloon car racing.

Ari Vatanen/Terry Harryman head for 4th place on the 1983 Acropolis Rally in the Opel Manta 400.

In 1978 Cosworth also took up the lure of rallying when they undertook to develop the Ascona 400 on behalf of Opel. The almost square Rekord diesel 2400cc was utilised as the basic item before receiving Cosworth internals for both road and competition models. Initially producing 240bhp at 7500rpm, the unit eventually saw over 270 with sufficient reliability to clinch the World Rally Championship for Walter Rohrl in 1981 after victories at Monte Carlo and the Ivory Coast. Next year Ari Vatanen took another Ascona 400 to win the Safari Rally. The Ascona was eventually pensioned off to be replaced by the mechanically similar Manta 400, but although well outclassed in the World series, the mixture was potent enough for Jimmy McRae and Russell Brookes to provide some epic dices *en route* to British Open titles. Cosworth's role in the KAA has been all the way from machining the castings supplied by Russelsheim through to servicing the engines.

There is virtually no aspect of major motor sport which has not benefited by involvement with Cosworth at some period, be it rallying or saloons, single seaters or sports cars. The competition would have been poorer, the entertainment lessened. The Company has always endeavoured to bring the same standards of quality control and craftsmanship to all their products and the unequalled list of events won and championships clinched is testimony to their resolve.

The 2400cc KAA engine was the rally version of the Opel Ascona 400.

When they pushed Martin Brundle's Tyrrell 012 away at the end of qualifying, it had not made the cut. That was Austria 1985 and *Wagen Drei* had narrowly failed to snatch the last slot on the grid.

Seven days later as the teams reassembled to do battle at Zandvoort, scene of the DFV's superb debut, more than a week had passed: 18 years. For the first time since 1967 a Grand Prix paddock would not resound to the throaty roar of a V8, Tyrrell, the last of the Ford Cosworth runners, having defected to the turbos. The team had been Northampton's first commercial customer after the Lotus exclusivity was lifted and its redoubtable leader had often expressed disapproval with the forced induction engines being allowed into his beloved Grands Prix. Now he had no choice if he wanted to remain in the sport as FISA had decreed that 1986 was to be for turbos only.

Another to be equally vociferous in his objections to the turbochargers was Keith Duckworth, feeling they were a distortion of the regulations. Supposedly cast in stone, the rules were more likely to be Plaster of Paris and easily moulded or broken to suit the sculptors - much to the delight of a certain local car manufacturer.

Nevertheless, a year before the final farewell of the three litre, Cosworth had come to an agreement with Ford to prepare a 1500cc turbocharged engine ready for Formula 1 in 1986. The initial plan was to use a BD-derived straight four cylinder, as Keith explains: "We had a go at a four then they changed the rules. There was no certainty as to whether it was to go fuel consumption or power. There was a time we thought it was going to be fuel controlled and a four would have been sensible." When the BD blew asunder on the test bed, unable to take Formula 1 boost levels, it marked the end of the road for that idea and thoughts immediately turned to a V6.

The design of the engine, known as the GBA for the Company's internal consumption, was entrusted to Geoff Goddard starting with a paper exercise back and forth between him and Duckworth to establish the fundamental parameters of the unit such as the still hush-hush bore/stroke dimensions.

Geoff describes his own notes as "Memos to the Chairman," while those coming his way were "Tablets from the Mountain."

Aiding Geoff in what was his first complete engine project, much as with his boss so many years before, were John Hancock and Roy Jones with the former detailing items like the block while the latter completed ancillaries etc.

Keith plays down his own efforts: "The only influence I had on it was the liner and fundamentals of block design, the fact that we made it 120°, what kind of balancing we would hang on it. Odd bits."

The 1.5 turbocharged engine developed for Ford to go Grand Prix racing again saw the first light of day mated to the Beatrice Lola THL2 for testing in February, 1986.

Designer Goddard continues: "You can never negate Keith's input, he is always in the background. His input was at the structural design stage, how to carry the loads around the engine, and then he came back right at the end with decisions of how he wished to see it mounted in the vehicle. The whole of the middle bit was left in this office so he might come back after six months and say we could have done this but by then it is too late!

"The basic layouts, the overall picture, naturally came first, then the cross bolting and various systems, piecing together the major components within the structural argument." Then comes the more detailed phases, essentially: "The bits that make a Cosworth a Cosworth, ensuring the detailing is up to the standards we expect." All the drive systems are then planned and then the working out where all the oil will go on the drawing board, "We hope the oil will understand the arrows!" Gradually, methodically, the GBA started to take on the shape that the world now knows.

Only about a dozen items were carried over from the DFV, things like a relief valve and the scavenge pump rotors, it really was a case of starting afresh. The layout was completed by October, 1984 so major components like cylinder heads were already into pattern making by the time Santa came a-calling with the rest of the umpteen pieces following until a completed engine appeared on a test bed at the beginning of August, 1985. It is interesting to note that none of this was achieved with computers, actually being done the good old-fashioned way, although Cosworth are evaluating the possibilities of such aids now that the cam aspects of cad-cam are closer to their precise needs. The highly skilled prototype shop, close at hand to the design office, do all the 'first offs' of any Cosworth project before 'productionising' occurs - a crankshaft perhaps taking 12 to 20 weeks due to the time constraints of heat treating.

After months of bench testing the V6 was finally mated to a chassis for track testing in February, 1986. The recipients were the new Beatrice team with their Lola THL2 for Alan Jones and Patrick Tambay, AJ having given the equipe their debut at Monza the previous September. The Hart 415T-powered device had lasted only six laps. It was to be an ominous note.

The first runs of the new combination were taken at Ford's private track of Boreham in the midst of an Essex winter, those assembled there wringing their hands for warmth as much as in silent prayer. The first thing to be noticed about the GBA was the overall compactness of the unit, substantially smaller than its intended competitors despite being a 120° Vee.

The first thing that the drivers noticed was a marked lack of oomph. Much of this centred around the highly complex engine management module coded EEC-IV from Ford's own electronic wizards. The wonder of aerospace technology controls a multitude of functions about engine performance and it would take a long time to 'map' the requirements correctly. Even at Adelaide for the last race of the '86 season there would be dramas and the ability to self-diagnose faults became a very useful faculty.

The GBA was a very compact unit and was substantially smaller than its competitors being a 120° Vee.

Another, more fundamental reason, was the simple fact that Cosworth had underestimated what the requirements would be for 1986 when they had sat down to plan the engine two years before. Not only had they planned for pump fuel as opposed to the special brews now operating in Formula 1, they had also totally miscalculated the equine potency of the opposition. The thoroughbred was in danger of becoming a mule.

Goddard recounts: "We obviously arrived at our best guess as to what it needed in the way of horsepower and found that the rest of the field went thataways! We came back, said 'That wasn't too bright' and turned the wick up." Cosworth had estimated on 650bhp. By the time they were ready for action the Hondas, BMWs etc. were well into the eights.

Jones gave the new combination its race debut at Imola at the end of April. It had been agreed from the start that there would be no special qualifying unit, no two lap demon, so it was no surprise when he lined up only 21st on the grid. This was to be no Zandvoort, only the naive and uninformed could ever have contemplated such a scenario. After making good progress until half distance the Aussie pulled off with overheating problems. His polite comments would become more strident as the season wore on, the mega-buck sponsor departed after a boardroom revolution of which the Formula 1 road show became a sacrificial lamb, team personnel were at each other's throats and finally two patrons pulled out. It was not the best environment in which to develop an engine.

The only high spots of the liaison were the encouraging pace at the 'Mickey Mouse' circuit in Hungary then bringing both cars home in the points at Austria. Sixth in Italy was the last 'hurrah' for the 'other' scarlet car.

Martin Walters, the Chief Development Engineer and man closest to the fray reflects: "The car went well at some circuits but at others, like Brands Hatch, it was an absolute disaster, undriveable."

It took some time to make up the leeway, mainly through an improved fuel system and changes to the compression ratio, while the cylinder heads remained unaltered since schemed in 1984.

Chief Engineer Duckworth: "I hung out the diameters and bearing sizes of the crankshaft and thought 'That will do' but was then only thinking of 650bhp. I thought it was going to be driven down to that whereas various rule changes not occuring and special fuels coming in meant it had gone up to 900bhp. So I look in awe that our little crank will take this without failing. I'm highly impressed."

The beneficiaries of all this progress are the Benetton team from Witney. Formed in 1981 as Toleman, then they were non-qualifiers but have improved out of all recognition to score their first Grand Epreuvre last season, down Mexico way. Optimism runs high. With drivers the calibre of Thierry Boutsen and Teo Fabi plus the renowned chassis genius of Rory Byrne, this amalgamation of talents will hopefully be the catalyst of success for the Ford V6 built by Cosworth.

After the dissolution of the Beatrice team at the end of 1986, it was the Benetton team who were the recipients of the 1.5 litre turbocharged engine in 1987.

Until Detroit the Benettons were a disappointment, but at Motor City the teams were given revised engines and turbos which delighted both drivers. For the remainder of the season they qualified well up the grid and were in the leading pack in most of the following races.

The Cosworth-designed engines in both the Sierra and the Benetton did a great deal to enhance Ford's reputation as the car company with one of the greatest sporting reputations.

When Cosworth first really set out to attract business away from the helter-skelter and capriciousness of motor sport, to encompass road cars within their sphere of influence, they were in for a nice shock.

There comes a point in the life of any racing engine, however glorious the past, where it no longer is a winner. This can be the result of technical betterment by the opposition or an arbitary change of rules to make the unit instantly redundant; and there is nothing so difficult to sell than a racing engine which no longer wins races.

As the eighties loomed then unfolded, Cosworth survived upon their ability to sell racing engines, little else. A major source of that income for many years had been the DFV, but now it was being forced out of Formula 1 by the screwdriver technology of the turbos. As the three litre market waned so the 2.65 waxed but the writing was writ large on the walls: the Company needed more stable areas for future growth.

Keith Duckworth explains: "We were always half interested in road engines. We had gone from being modifiers to manufacturers of racing engines so were forced into the top echelons of motor sport.

"As these showed instability it did not make it look commercially viable so we decided to sell conversions to a road engine as well as our design knowledge. We had developed good numerical control methods of machining heads that were right for small production runs so it was reasonable that we should make the heads for image-performance road cars. With racing having gotten so expensive it would seem a sensible direction to expand the firm in. Our knowledge gained in racing can be usefully sold into production cars because although there are still great differences you still have to win. The training and basic principles are the same."

The first major response to the quest was when they had a 'jahr' from a very unexpected source: Mercedes-Benz. The second surprise was that they wanted a competition engine!

The most powerful model in the Mercedes-Benz 190 range was the 2.3 litre four-cylinder 16 valve version with the Cosworth head.

Clad with glassfibre side skirts, aprons and a prominent spoiler on the boot lid, the 2.3-16 was an obvious performance car.

The German firm are renowned in the motor industry for their exacting standards with quality control second to none. Any supplier must be of the very highest calibre with exemplary products if they were to pass the stringent examinations of Stuttgart. To give such a commission to a foreign company was probably unprecedented. More formally it is known at Cosworth as the WAA while the car giant designates the product as M 102. The final outcome is known by the world at large by the liberal name of Mercedes-Benz 190E 2.3-16.

Mike Hall is now a director of Cosworth and in charge of the road engine designs of the Company. "They originally commissioned us to do a Group B competition engine, and that is what I designed, based on their 2.3 litre block. It was dry sumped, fuel injected, all-can-do race/rally. The prototypes were built and development carried out at Stuttgart."

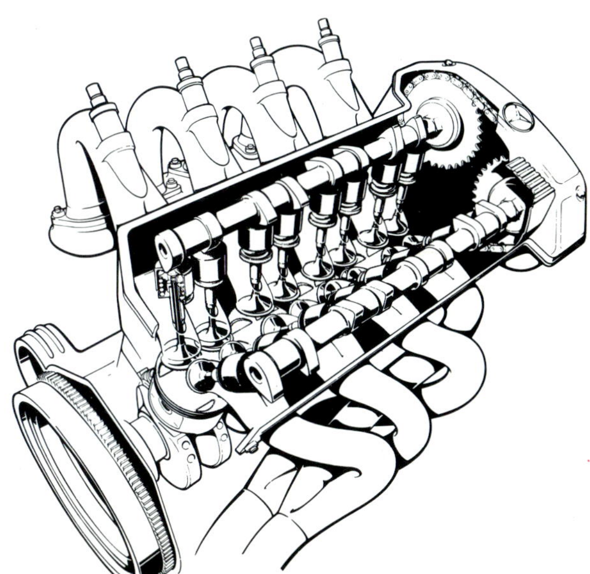

The 2.3-16's block comes straight from the standard 190 and is crowned with the light alloy, twin overhead cam Cosworth head.

Then came word that the Germans were to drop the project, bad news. Instead they required engines for a production run! Hall again: "It was good news for us as we were only involved in making about 200 for homologation," now there would be thousands. "A few modifications took place on the design: different camshafts, ports made to suit the torque characteristics of a production engine, valve springs, they really were the only things we had to alter. Now we were only making the cylinder head rather than a complete dry sumped racing engine. In rally trim it was giving 320bhp very early on so everything is there should Mercedes ever reinstate the programme."

To date the factory has not taken up the challenge although three world records did actually fall to the 190E during extensive endurance testing at Italy's famous Nardo track in 1983. Others, however, have not been able to resist the temptation of having that all too rare honour of competing with a 'Merc', notably the likes of Danny Snobeck and Alain Cudini in France plus a variety of names including Ian Taylor in some British based racers.

Cosworth took up the challenge of producing cylinder heads at numbers which far outstripped anything they had previously done, and they were doing it for the most exacting of task masters.

The WAA project not only set up whole new targets and opportunities for Cosworth, it also created engineering firsts for the Company like their integrated cam carrier/cylinder head which was more efficient and easier once tooling up had been done. The result of these labours is an extremely quick and refined luxury express. Its straight four engine has twin overhead belt driven camshafts, very much in the Cosworth family tradition, giving 185bhp at 6200rpm from 2299cc. One nice touch during manufacture is that the cylinder heads are hand finished prior to assembly at Mercedes' specific request. Cosworth consider their standards are so good as not to require this but if it is what the customer wants....

Northampton thought they knew what the customer wanted when they grafted one of their special cylinder heads onto a two litre Ford Sierra. The result was 160bhp of a very quick motor car certainly capable of breaking the jellymould. They were wrong, Ford wanted more, they wanted a Group A racer. To be accepted for the category they had to sell a minimum of 5000 similar vehicles in 12 months. It was another mountain for Cosworth to climb.

The Ford Sierra Cosworth engine production line at Cosworth's factory in Wellingborough.

By winning the Tudor Webasto Manx International Rally in September, 1987 in his Ford Sierra RS Cosworth, Jimmy McRae won the Shell Oils RAC British Open Rally Championship and clinched the Manufacturers' Award for Ford. With 20 outright International rally victories to its credit by mid-September in its first full season of competition, the Ford Sierra RS Cosworth had already secured more victories than did the world famous Ford Escort rally car in its debut year.

To possess a potential winner, Ford deemed a minimum of 200bhp was needed from the road going derivative from which the competition models would evolve with scope for 300 plus. Cosworth delivered 204 at 6000rpm, a Garrett T03 making the difference. Allied to a T88 'Pinto' block with fuel injection and the latest in fuel management systems it is a relatively simple matter of altering the boost pressure and inserting a revised microchip into the electronic brain, then, hey presto, its competition time! With sporting considerations very much to the fore the YB-series of engines had already progressed into the alphabet with -C for rallying and -F to race. YBE has been temporarily excluded from the list after some wag from Ford suggested it might stand for 'Your Bloody Engine'.

The Ford Sierra RS Cosworth has styling as impressive as its name. Since being launched it has become *the* vehicle for those who want to be ahead of the crowd. It applies also to competition where in rallying, the forests are full of them while saloon car series are awash with them. Why be slow? As with FJ, SCA, DFV or many another category, if you want to stay ahead, get a 'Cossie'.

To facilitate the production of all these road engines Cosworth started to look for suitable premises to undertake the volume. They subsequently acquired a new factory at Wellingborough in 1984 totalling just under 30,000 square feet.

It takes less than three hours to machine a bare Worcester casting into a complete cylinder head ready for action. This is done on four Heller computerised work stations where up to 160 different cutting tools can be utilised on each machine. A fifth Heller has been especially modified by Ben Rood and his team to ream the valve guides and cut their seats, this masterpiece of innovation being achieved at a fraction the cost of purchasing a machine for the specific task.

Sierra by Cosworth out of S

Thank you Cosworth for another great engine.

Some Mercedes head castings used to go to Stuttgart for machining and assembly but now Cosworth does them all. The Sierra, however, is not yet finished because Cosworth actually assemble the complete engine before despatch, another first for the Company. They procure all items from washers to blocks to average 22 finished engines per day on current schedules. The total capacity under Wellingborough's roof is around 10,000 per annum so further expansion appears inevitable.

Then, who knows? The initial 5000 Sierra engines have long since gone, work continuing on competition and evolution types of a contract still with many years to run. Next to receive the Cosworth treatment is the Sierra Sapphire, so giving another dimension to the expression 'to boot it'. Ford, for their part, have already expressed interest in the possibility of having Cosworth interpretations of other models to compliment the Ghia luxury versions.

Keith Duckworth believes it is a burgeoning sector: "I believe the cost of manufacturing engines on transfer lines, the number you have to make and the competition within the motor industry means that the majority of road cars will become same-ish. Therefore, there will be a growing market for niche cars so that people can have something different."

A growing number of leading motor manufacturers are heeding the words and following the path laid down by Mercedes and Ford in search of eldorado, Cosworth's golden touch. Projects still embargoed are already taking shape for other manufacturers keen to exploit this unsurpassed knowledge and experience in high performance engines. Many prefer to have the association secret, preferring to keep all the kudos for themselves whereas others such as Jaguar are happy to announce their links, two great names sharing a common bond, the same goal.

An F-type Cosworth. Wouldn't that be a nice shock!

The Ford Sierra RS500 Cosworth was an evolution model of the Sierra RS Cosworth which could develop 224 bhp for road use compared with the 204 bhp in the regular Cosworth, but it had the potential to exceed 400 bhp in race trim.

Any Other Business was the category under which the Ford hierarchy first discussed the DFV proposals. When applied directly to Cosworth's own business interests, it is a title going beyond dramatic roadsters and chequered flags, for many are the aspects of company involvement that go unheralded by public acclaim.

It is possible that on a clear still day the melodic tones of Worcester Cathedral's renowned choir may drift across the waters of the River Severn into an unpretentious factory unit on the opposite bank, slightly upstream, but once inside, the crescendoes are of a different fire and brimstone because herein lies the Cosworth foundry.

As the seventies progressed Cosworth Engineering encountered severe problems with the castings they purchased from commercial foundries - cracking, distortion, high scrap rates etc. - which affected not only machining and reliability, but also inhibited development of their racing engines. Protestations went unheeded because the foundries were unwilling to change the ways of yore, preferring to rely on the black arts and little science that had served them all well since Adam was a lad. Also, to an industry who traditionally sell by the ton, the quantities required by a small Northampton firm were insignificant, so why bother? Take it or leave it.

They chose the latter. A noted metallurgist and physicist, named John Campbell, was recruited to conduct a detailed survey and it was to nobody's surprise when his conclusions indicated that existing practices were inherently flawed so only a substantially new approach would achieve the standards Cosworth demanded. They had no choice but to go it alone.

The riverside plant was acquired in 1978 under the banner of Cosworth Research and Development Limited to derive an improved process to produce castings for the parent company in Northampton. Within a year the principles and basic techniques of the process had evolved with castings of higher integrity than any they had bought in being made by their own capable hands.

Bob Smith, Managing Director of the facility, takes up the story: "The Cosworth Process is, firstly, the preparation of clean metal. We do not use reclaimed alloys, only high purity metals like LM25 although the process can use a whole range if required. Secondly we put it into a mould that keeps it clean because contamination of the metal leads to most of the structural problems. Thirdly, we use Zircon sand moulds actually to make the shapes (Zircon is a rare, expensive material acquired from Australia. It can be used almost indefinitely, unlike conventional sands, and has many desirable properties including better consistency). 3A is the reclamation of the sand to make totally economic use of it."

The process draws from the centre of the molten mass thereby allowing impurities either to sink or swim in the furnace before gently filling the mould from below by way of a computerised electro-magnetic pump borrowed from the nuclear power industry. This eliminates turbulence so avoiding oxidisation and porosity, *bête noire* of old-fashioned methods. It all seems so simple!

One of the first prototype castings from Cosworth's foundry.

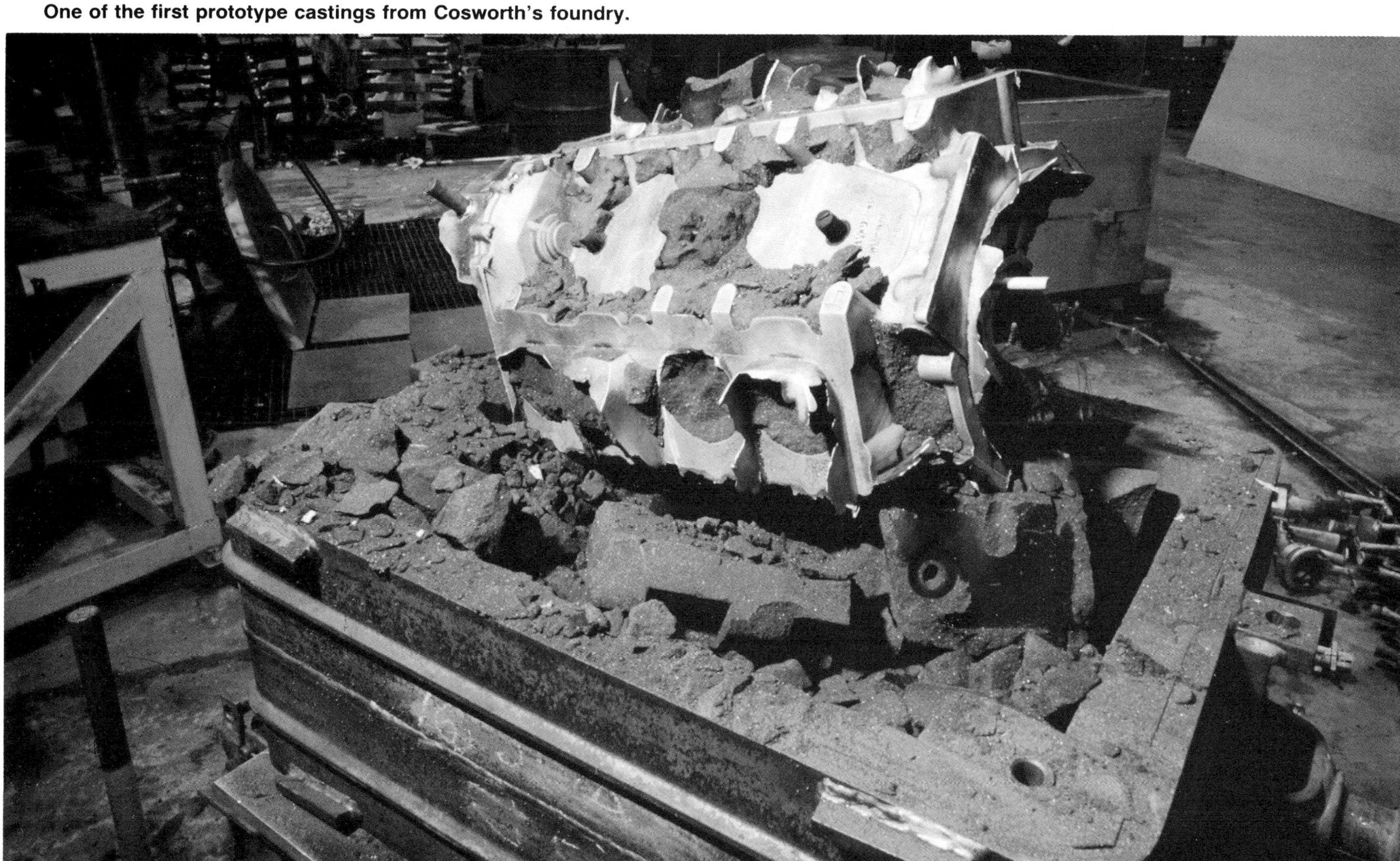

The Ford Sierra RS Cosworth was the first car bearing the name of both companies after 25 years of collaboration. There were a total of 92 detail bodywork modifications incorporated into the standard Sierra's sheet metal. The large rear aerofoil generated up to 45lb of positive downforce at around 70 mph, while other main additions to the bodyshell included wheel arches, sill extensions and a deeper front air dam.

Some products of the Cosworth factory (below left) and one of the very early closed cycle diesel engine test rigs (below right).

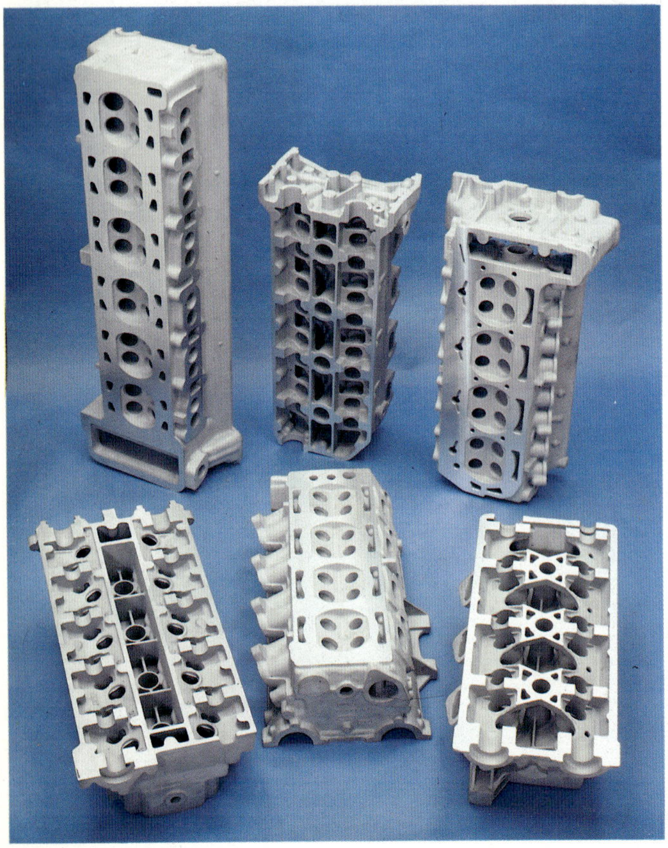

Nowadays, the Cosworth Process is patented and marketable with a number of enlightened sources expressing interest in building similar facilities under licence. Very recent developments will speed up the process many fold thereby greatly enhancing its commercial viability.

Bob Smith: "It was a moribund industry, seen as low tech, that was it. For a number of years Cosworth was regarded with interest, but some cynicism. We now produce the finest commercial castings there are, in particular, complex items like four valve integral cam-carrier cylinder heads.In terms of accuracy, they cannot be beaten. I do not think anybody can equal them on a consistent basis. They certainly cannot beat them anywhere in the world. The new ideas dramatically transform the economics and we believe it can only be done by the Cosworth Process so we now have the ability not only to make the best but also the cheapest castings!"

Bold words indeed but the men of Cosworth are used to making winners and the sentiments are backed up by some fascinating observations. Every 'four pot' cylinder head, be it for Mercedes or Monza, must be accurate to within a MAXIMUM tolerance of plus/minus twenty thou (0.020") in any plane - a good machining target let alone a cast straight from the mould - with production averages considerably finer. Likewise, combustion chambers are to within plus/minus six thou (0.006") as cast.

Comparing a DFV head from 1977 with a typical Worcester product, it is amazing to see the former is a whopping 5m/m oversized while the foundry's own is a mere ½mm - both nominally from the same drawing! This improvement also channels into a number of others including controlled wall thicknesses for waterways and better heat dissipation. It is a quantum leap forward in integrity that probably needs to be seen to appreciate fully, but as the Managing Director is proud to say: "It is a work of art, but not one that required artistry, only science, to produce."

The initial premise was to provide castings solely for Northampton. Today the two foundries service internal requirements and a burgeoning order book of the subsequently formed Cosworth Castings. The big commercial breakthrough came with the Mercedes 190 competition engines Mike Hall was to design, so when this escalated into the thousands necessary for production quotas, Worcester was initially overwhelmed, both by the prestige and the volume, until the second system was fully operational. Now they produce many thousands of cylinder heads per year plus a myriad of other components, many containing features that others would need to machine-in later. The plant is capable of tripling its output, even without the new high speed mods.

Amongst the motor industry hardware made there have been the heads for the Metro 6R4 rally effort while over 2000 complex inlet manifolds are created each year for Rolls-Royce. Also bearing the famous name is the air intake for Lynx and Sea King helicopters of which Cosworth are particularly proud. This is, probably, their *piece de resistance* for they are the only company licenced by the manufacturer to make this vital, highly stressed, item. "It is a 32 X-ray shot, Class I aerospace casting. We have supplied 900/1000 and none has ever been failed for casting reasons. We have never had one returned."

None of the academics or other interested parties who regularly analyse a variety of Cosworth castings have ever complained either although such high standards are not without their problems, namely being too good! Why? "Because of the accuracy we achieve people have to re-tool if they want an existing component made by us. The European defence market is very small so they don't want to spend the money."

Such are the exacting standards at Cosworth that every 'four pot' cylinder head must be accurate to a maximum tolerance of plus/minus 0.020″ in any plane.

Notwithstanding, the castings are pushing engineering excellence to new highs through being superior to anything previously available, be it a cylinder head for a new performance car or the block of a two stroke V6 outboard motor, the high pressure in-flight refuelling manifolds for a VC10, gun cradles required by the Royal Ordnance Factory or hull sections on a submersible minesweeper.

Bob Smith and his team know that they are the best. As he says: "We just have to go out and evangelise." Across the waters from Worcester Cathedral.

If the foundry was born out of frustration then it only mirrors the story apropos pistons. They may go up and down but in the early seventies Cosworth Engineering were going round and round in circles trying to resolve problems such as poor delivery from their existing suppliers. Consequently it was decided to make their own and in 1974 Jeff Roper returned to his previous employer to conclude the purchase of a forge when Hepworth & Grandage decided to cease forged piston manufacture. The machinery was acquired and installed for a fraction of its new cost into a new building directly opposite headquarters at 'Jimmy's End'.

Initial production was for in-house requirements only but soon the enquiries started to flow in, first from Chevrolet, to become a tide which could not be turned. Now over two-thirds of all are for outside contracts. Large or small, road or race, to date Jeff Roper has designed over 1000 different pistons and built up a stock of over 70 dies - up to 118mm diameter - some of which have been the basis of dozens of uses before machining turned them into individual components to fill specific orders. The Toyota Formula 3 race unit, for example, shares a common die with the old DFV 3 litre engine.

Output is 25000 per annum with about one-eighth devoted to their own DFX programme while the main exterior customers are Aston Martin and Chevrolet. The average batch is 75 although if you have a masterpiece hidden away in a barn, *hors de combat* for want of an engine rebuild, Cosworth can probably supply a minimum order of two dozen to help you on your way, be it Mustang, Piper or *Cavallino Rampante*....

A trip down memory lane via the Roper logbook recalls much splendid machinery and many a famous name. Owners of beloved ERAs and Maseratis, BRM V8s and Climax FWAs are amongst the recipients of Cosworth's finest, as are restorers/renovators like Morntane, Maranello and Graypaul. Contrasting classics such as a supercharged Alfa straight eight, a Ferrari 330 P4 and John Fitzpatrick's Porsche 956 share a commonality with Matchless G50, Manx Norton (a 'real lump' according to Jeff) and Elmer Trett's Kawasaki dragbike - 0 to 200mph in seven seconds is a great test of a piston! - representing the two wheeled brigade. In Rover V8s, in differing sizes, for the TVR 4.2 litre road burner or TWR's 180mph Group A race winner. They are also found in Jaguars and those majestic 'Silk Cats' for *Les Vingt Quatre Heures du Mans* - 630bhp, 7000rpm, 24 hours....

Cosworth have even supplied their nominal Formula 1 rivals, ostensibly the Matra and Autodelta twelves of a decade ago, although today such approaches are refuted, as Jeff explains: "Keith has tended to put the kybosh on it." Succinctly put.

The list continues to span the globe: Weslake, Ford Boreham, Ryan Falconer, McLaren USA, Dealer Team Holden and Toyota Team Europe while mention of the Japanese highlights the fact that Northampton now supply the majority of 'works' supported sports car teams who threaten a new dawn of competition from the land of the Rising Sun. In the United States both Buick and Pontiac carry the British component as do SCCA racers. One Good Ol' Boy is known to have purchased Cosworth pistons despite being *given* a local brand by a sponsor, while many are the NASCAR heroes who propel two tons of metal high up 'in the groove' on the Daytona banking.

Amongst all these applications it is ironic that the car bearing the company name, the Sierra RS Cosworth, contains someone else's pistons! Unfortunately the division cannot cope with the numbers required for such a project so, with hand on heart, the internals are provided by Mahle to strict Northampton specifications. The competition versions, with their limited numbers and higher demands, carry the Cosworth product.

The Sierra story highlights the deliberate policy not to allow piston production to grow too big, rejecting the temptation to expand and risk loss of quality. Likewise there is no compromise on basic material because although investigations do continue into things like chrome-zirconium, until they prove superior, then Cosworth will continue to use only the best currently available which is specification RR58, the same alloy as powered the Supermarine Spitfire in its daring deeds, and will continue to fly high as Britain's premier specialist piston manufacturer.

If you have a piston you also need a connecting rod.

Since about 1983 Cosworth have been infiltrating the conrod market in America, primarily aimed at Chevrolet racing engines. Nowadays they make about 4000 items per annum to include both Pontiac and Oldsmobile engines as well. They are fashioned from a single rod forging and single cap forging with one type of bolt, a ⅜". Interestingly all American made conrods feature a 7/16" bolt but Cosworths have never had any problems with their smaller variety, an excellent indication of the product's quality. All are made from the best quality steel, and to the same high standards as those in the 'Indycar' DFX engine. Eventually it is hoped to be able to cover a much wider range.

Companies like the Arrows Grand Prix team commission high precision work on complicated subjects like the magnesium casing of a Formula 1 gearbox while Ford at Halewood required the mainshaft for those overdrive transmissions as fitted to export models of the Transit van. The cylinder heads of the Chevette HS2300s were machined from castings supplied by Vauxhall while the Austin Rover Group got Cosworth's own foundry to make the heads for their 6R4s before sending them on to Northampton for machining and assembly.

Big prize or big load, it is all the same to Cosworth Engineering.

With the increasing usage of electronics in engine management, the Company has an expanding section to deal with the chips and bytes. Both fuel and ignition systems are catered for as 21st century software is mated to 19th century internal combustion hardware.

All the high-tech contrasts obliquely with a hidden corner of the prestigious Wellingborough plant, where beyond the masterpieces in metal hides a novel collection of hardware. It is what site director Harvey Fox calls: "A closed cycle diesel engine which will work underwater at virtually the same efficiency as it would on the surface," and can be regarded as his hobby. "We take a standard diesel engine, scrub out the CO2 from the exhaust, which goes back into the intake with the amount of oxygen that is burnt during the cycle and a small amount of a virtually insoluble inert gas to restore the ratio of specific heats to that of air." As with everybody at Cosworth, he has the ability to make the complex sound matter-of-fact.

The project started about four years ago as the result of a public plea by Prof. Thompson of Newcastle University for industrial involvement with his project. Cosworth were looking at ways to diversify, "to take away the peaks and troughs of racing because they tend to change the rules as they like." So after both Messrs. Fox and Duckworth had studied the potential - aided by Keith's soda syphon - it was agreed to go ahead.

Former Rolls-Royce technical chief Fred Morley was seconded to the project and now there is a team of six 'Royce' pensioners working to prove that you are never too old to learn.

Harvey notes: "The only problem when we meet up once a fortnight is that they do their drawings in imperial, and they all tend to shout at each other!"

Such are the consequences of 'Run Silent, Run Deep' with sufficient progress made to get two prototypes into the metal for stringent testing with prospective clients. The top man continues: "The total auxiliary power required is about 15 Kw for something in the region of 120/150 Kw. It is a very efficient system that will work at any depth. The two devices are for submarines. It could be used in torpedoes, trenching, laying pipes, oil rig inspection, mines, anywhere the environment is susceptible to dangers. There are, literally, scores of possibilities."

Although regarded as revolutionary, now the basic design has been finalised, it is not seen as significantly difficult to manufacture. The most critical parts, sleeves and valves, have been produced, quite naturally by Cosworth themselves, with the bulk of the remaining 500 components contracted out. Although the prototypes are clad in stainless steel, production examples will probably require more exotic materials to combat corrosion etc. while in service. Internally some items are made with tungsten carbide - experiments to fill it with sand failing to impair as yet!

Much work still needs to be done before the project becomes commercially acceptable, so marketing remains passive while: "This is never going to become mass production, maybe 10 a year."

Enthusiasm for Cosworth's strive for perfection permeates all aspects of the factory.

"We are enthusiastic, not enthusiasts," is how senior employee Bill Pratt describes the Cosworth approach to life in the fast lane.

Enthusiasm permeates all aspects of the Company: from Gloria as receptionist/ switchboard operator - most people's first contact with the firm - to the Chairman/chief engineer at his loquacious best. They have the confidence of champions, the surety of success.

The empire founded by Messrs. Costin and Duckworth has become prosperous far beyond the original dreams 'to make an interesting living from racing cars and engines'. Unable to rub two Coventry Climax washers together, it is now a multi million pound company employing nearly 500 people. It is a far cry from the Christmas of '59 when Keith was refused a £30 overdraft to pay the telephone bill....

The enthusiasm, though, is not fiscal, it is forged in steel and cast in aluminium like the creations bearing their name. It is knowing that Cosworth, in all its guises, produces the best castings, the best pistons, the best engines anywhere in the world -bar none. The standards of design, engineering and production are unequalled. Consequently their services are in high demand at all times by the very best top names in motor sport, the automotive industry, aerospace and defence specialists. When it comes to a difficult, delicate task requiring the utmost skill and quality - if anybody can, Cosworth can. Hence the helicopter air intake for Rolls-Royce, the cylinder heads for Mercedes, the unsurpassed DFV/DFX roll call of honours. Nobody does it better....

These days Keith Duckworth has to combine his time as chief engineer with the duties of chairman - one look at the sectioned components and technical drawings strewn about his office gives ample indication as to which he prefers, where his heart still lies. For a man to have so much but still seek the infinite, the ultimate engine, is beyond the comprehension of humble scribes.

Mike Costin heads the Cosworth Engines group with overall responsibility for the fast developing road aspects of the Company, "which is supposed to be devoted to production engines, notwithstanding the fact that our last project was a racing one!"*

Old habits die hard. This man's role in the success of Cosworth cannot be overestimated.

Nor can that of Ben Rood, as good a production engineer as can be found anywhere. Ben's current tasks within the Company are both unique and vitally important: "Special processes and machine tools that we design and build here for internal use.

"We build them because we have a requirement for a machine and it is unobtainable or only at astronomic cost. We have built some that are unique in their fields, all developed for our own use to help make things in a very efficient manner."

The list includes conrod makers, block line borers and that Wellingborough gem to cut valve seats in the cylinder heads - some 30 or 40 in all. A design is only as good as the ability to manufacture it properly and Ben's often unheralded contribution has been to ensure that Cosworth get the very highest standards.

The fact that the three men have remained good friends for so long is a credit to all concerned while the lack of 'bull' instigated by clear thinking engineers reflects even in the Company rule of addressing everyone irrespective of rank, by their Christian name.

There is also not for Cosworth the ritual of an Executive Dining Room: if the boss does not bring his own lunch, it is into the canteen with everyone else. The bluff, no nonsense approach is reflected in their business dealings, as Keith remarks: "I am interested in understanding things that will stand rigorous analysis. We are terrible to people who try to sell us things. We ask them why and how and if....Normally they haven't a clue! I gave up consulting experts on that basis in about 1963!"

One thing Duckworth and his partners do understand is the continuance for the future. They have encouraged thinking, knowledge and skills within the organisation with young apprentices taught a variety of tasks so as to give them a well-rounded education in the best that engineering can offer. Cosworth training has proved to be highly desirable and renowned with the likes of Brian Hart, Alan Peck and Terry Hoyle all benefitting enormously from their experiences on the Cosworth payroll.

To further ensure the long term stability, in 1980 Duckworth sold a large portion of his shareholding to U.E.I. The entrepreneurial technological and electronics group has an annual turnover exceeding £100 million, its most famous enterprise - besides Cosworth - probably being Quantel whose 'Paintbox' and 'Mirage' digital techniques produce all those bent, bowed and wrap-around images beloved by the 'Beeb'. If you believe the camera never lies you have never seen Quantel. Cosworth Engineering's independence as a

The U.E.I. Quantel Mirage system produces all those bent, bowed and wrap-around images so often seen on television nowadays.

*The DBA, a D.O.H.C. 16 valve version of the famous Pontiac 'Iron Duke' engine having its first test-bed runs during the writing of this manuscript.

nanufacturing company remains unaltered y the liaison.

Also proud of their independence are the wo off-shoots at Cosworth-upon-Severn. The oundry maintains strict confidentiality over nany items for commercial reasons, North-mpton often not aware of the goings-on in Vorcester. As far as the boys from the otstuff are concerned, Northants is a cus-omer, albeit a privileged one.

Wellingborough continues to grow towards ull capacity and as other projects come on tream will undoubtedly need more space. If ne submarine engine ever gets off the round, for want of a better phrase, then....

Space is also at a premium in St James Aill Road where around 80,000 sq. ft. is lready owned by Cosworth with leases on ther property. This reflects the continued uccess of the Company which in 1987 has rders for more V8s than at any time in its istory.Not bad for a 20 year old engine which vas only expected to remain competitive for iree or four years.

Nowadays, Northampton is more than)FVs and DFXs, it is the hub of the greatest pecialist automotive (and much more be-ides) engineers in the world. If the Govern-nent wanted a prime example of Britain at its est they need look no further. Indeed the Cosworth envelope has received a number of wards, notably Queen's Awards for exports nd technical achievement, while it is now)avid Keith Duckworth O.B.E.

Brian Hart has recently come back into the old when Cosworth took a majority interest in is Harlow based company. Employing about 4 persons, Brian's prime rôle initially will be service the Tyrrell DFZs. The agreement llows him still to pursue his own interests vith those successful 420Rs and should nyone wish to resurrect the Hart 1.5 litre urbo....

Abroard the subsidiary company in Califor-ia has about 7000 sq. ft. for its dozen mployees where substantial spares are held n stock for American customers and some)FX rebuilds also occur. Cosworth used to cence a number of companies to rebuild)FVs in Europe, including Engine Develop-nents in Rugby where John Judd built a umber of special engines for those superfast Villiams with which Jones, Reutemann and Rosberg battled so bravely against the turbos few seasons ago. Another is Heini Mader in Switzerland who is known to have reflected n the fact that in the language of those Gnomes of Zurich, the DFV is the same price ow as it was in 1967/68, about SF90,000. Such is progress.

There are people who rebuild other types f engines too, there are suppliers of the very op grades of steel and alloys, heat treatment specialists etc. etc. All have Cosworth to thank for a major contribution to their liveli-hood. There are pistons and pilots, races to be run, roads to be travelled....

Three decades on, the pair of likely lads have come a long way, they have fulfilled their ambition and more.

It is still 4000 miles from Indianapolis though. Still denied entrance to 'Gasoline Alley' by the redoubtable official, Malcolm Tyrrell and Keith Duckworth were perplexed as to how they could gain their rightful access.

Coming their way aboard a 'monkey bike' the pair espied Al Unser, the first man ever to race, the first man ever to win with a DFX. Salvation was at hand, or was it?

Imploring Unser to vouch for them, Al shrugged, stated that he had never seen either of them before and went merrily on his way, laughing.

Returning some minutes later the Cos-worth men were still without. Unser stopped and told the official to let them in. He did so with the comment: "If he says you are okay, you're okay."

As Unser sped off aboard his machine he turned, smiled and called to Keith: "Duck-worth, you owe me one."

Quite the contrary. It is he, it is we who 'owe one' to Keith Duckworth and to Cos-worth.

Keith Duckworth, Mike Costin and Ben Rood, the founding directors.

HISTORY OF ENGINES MANUFACTURED BY COSWORTH ENGINEERING LTD.

MAY 1987

Year	Engine	C.C.	BHP	RPM	Remarks
1959	MK I	997	-	-	The first modified 4 cylinder pushrod Ford 105E Anglia engine. An experimental engine not made for sale.
1960	Mk II	997	75	7500	The first production racing engine. A development of the Mk I and used in Formula Junior. Used A2 profiled camshaft.
1960	Mk III	997	85/90	7500	An improved version of the Mk II using A3 profile cam with a strengthened bottom end and (optional) dry sump lubrication.
1961	Mk IV	1098	90/95	7750	A bored out version of the Mk III fitted with bigger valves inter alia.
1962	Mk V	1340	80	5000	An inexpensive modified 4 cylinder pushrod Ford 109E Classic engine made for Lotus for installation in the Lotus 7. It had few modifications since it was intended only for road use.
1962	Mk VI	1340	105	7000	A racing version of the Mk V. Very few were made.
1962	Mk VII	1475	120	7000	A bored out version of the Mk VI. Very few made.
1963	Mk VIII	1498	90	6000	An inexpensive modified 4 cylinder pushrod Ford 116E Cortina engine produced for the road going Lotus Super 7 which replaced the Mk V. Fitted with an A2 profiled camshaft, 5 main bearing crankshaft and twin 40 D.C.O.E. carburettors. A large number were produced.
1963	Mk IX	1498	125	7000	A racing version of the Mk VIII.
1963	Mk X	1498	-	-	Purely a one-off experimental engine, this was the forerunner of the production Lotus Twin Cam engine as fitted to the Elan. The production engines, however, of 4 cylinders, had a capacity of 1558cc utilising twin overhead cam shafts (chain driven) with a cross flow cylinder head.
1963	Mk XI	1098	110	8000	An updated version of the Mk IV Formula Junior engine but with a very much improved specification. A very successful engine and a major development in Cosworth's history.
1963	Mk XII	1594	140	6500	A bored out racing version of the Lotus Twin Cam engine with dry sump lubrication though using standard connecting rods and cast iron crankshaft.
1963	Mk XIII	1594	140	6500	A development of the Mk XII with improved dry sump lubrication and latterly using steel connecting rods and crankshaft.
1963	Mk XIV	1498	100	6000	Another modified Ford 116E engine made for the Lotus Super 7 but with more power than the Mk VIII.
1963	Mk XV	1594	140	6500	A wet sump racing Lotus Twin Cam engine fitted to the Lotus Cortina for (then) Group 2 saloon car racing.
1963	Mk XVI	1498	140	6500	An 81mm bore version of the Mk XIII. It was very popular with the (then) current Formula in New Zealand and Australia to where almost all these engines were exported.
1964	Mk XVII	1098	120	8500	A downdraught version of the Mk XI engine with an improved crank design. Few were made, partly due to the brazing problems involved in modifying the cylinder head to a downdraught configuration. More a development rather than a production project.
1964	S.C.A.	997	115	8750	S.C.A. = single camshaft type A. The first extensively Cosworth designed engine, based on the Ford 116E Cortina cylinder block. (The block was the only "standard" component). Used in Formula 2 it had a gear driven single overhead camshaft, 5 main bearing crankshaft and a bowl-in-piston combustion chamber. A very successful engine which won the Formula 2 Championship in 1964 and 1965.
1964	S.C.B.	1498	175	8000	A 1500cc type B version of the S.C.A. made for experimental purposes only.
1965	S.C.C.	1098	135	9500	Same as the S.C.A. but with increased bore. It was destined for a North American sports car Formula but only a few were made. Many S.C.A. engines, however, were later sold to the USA and converted to S.C.C. specification.
1965	M.A.E.	997	100		M.A.E. = Modified Anglia Series E (Ford 105E). A very successful Formula 3 engine made in large quantities. A modified cylinder head introduced in 1966 improved the performance slightly but after 1965, few complete engines were made. Many kits and parts, however, were sold to other engine specialists throughout the duration of the 1000cc Formula 3 regulations.

Year	Engine	C.C.	BHP	RPM	Remarks
1967	F.V.A.	1598	218	9000	F.V.A. = Four Valve Type A. An unbeaten Formula 2 engine, winning all Formula 2 Championships from 1967 to 1971 - the duration of the 1600cc Formula 2 regulations. Based on the Ford 116E Cortina block, the engine featured 4 valves per cylinder with gear driven twin overhead camshafts.
1967	F.V.B.	1500	200	9000	An experimental 1500cc version of the F.V.A. made to examine the power output possibilities for the D.F.V.
1967	D.F.V.	2993	405	9000	D.F.V. = Double Four Valve. Cosworth's first Formula 1 engine and most successful engine ever used in Grand Prix racing. First fitted to Jim Clark's Lotus 49 at the Dutch Grand Prix in 1967 which it won. Since then it has won 155 Grands Prix and 12 World Championships. A lightweight V8 with 4 valves per cylinder and first Cosworth designed cylinder block.
1968	D.F.W.	2491	358	9750	A successful 54.0mm short stroke version of the D.F.V. made for the Tasman Series.
1968	B.D.A.	1601	120	6500	B.D.A. = Belt driven type A. A roadgoing engine designed for Ford for installation in the RS 1600 Escort. Using the 1600cc cross-flow Ford block, its design is based on the F.V.A. but with toothed belt drive to the twin overhead camshafts. 4 valves per cylinder.
1969	F.V.C.	1790	235	8750	A long stroke version of the F.V.A. made mainly for the European 2 litre sportscar championship which it won twice.
1970	B.D.B.	1700	200	8250	A development engine made for Ford for rallying application. Modified B.D.A. with enlarged bore. Only 1 made by Cosworth but many sold in kit form.
1970	B.D.C.	1700	230	9000	A fuel injected racing version of the B.D.B. developed mainly for Group 2 Saloon Car racing. Only 1 engine made by Cosworth but many sold in kit form.
1971	B.D.D.	1600	200	8750	Formula Atlantic version of the B.D.A. Only 1 engine made for development but many sold as kits.
1972	B.D.E.	1790	245	9000	A Formula 2 version of the B.D.A. bored to 85.6mm.
1972	B.D.F.	1927	270	9250	The B.D.F. improved on the B.D.E. by coming closer to the 2000cc Formula 2 engine limit although its 88.9mm bore was achieved only through brazing liners into the standard block. It won more Formula 2 races than any other engine in '72.
1972	E.A.A.	1995	275	9000	Based on the Chevrolet Vega block - Cosworth's first non-Ford based stock-block engine. An all aluminium 4 cylinder engine which was similar to the B.D.A. in layout and also had belt drive. Designed for Sportscar and Formula 2 racing but was never used in Formula 2.
1973	B.D.G.	1975	275	9250	An improved version of the B.D.F. but with greater reliability. The 1973 Formula 2 engine.
1973	F.V.D.	1975	275	9000	A modified version of the F.V.A. made for experimental purposes only.
1973	G.A.A.	3412	440	8750	A Ford V6 block based engine for use in Group 2 Saloon Car racing. (G.A.A. = G.A. engine type A, G.A. being merely letters in a new alphabetical code rather than a description of the engines.
1973	B.D.H.	1300	190	10000	A small capacity version of the B.D.A., based on the Ford 1300 cross-flow block. Group 2 Saloon Car racing and sportscars.
1974	J.A.A.	750	65	7000	A production motor cycle engine designed and built for Norton Villiers.
1974	B.D.J.	1098	150	9500	Designed for American S.C.C.A. Formula 'C'.
1975	J.A.B.	750	95	9750	Racing version of the J.A.A.
1975	B.D.M.	1599	225	9250	Large Valve Fuel Injected version of the B.D.D.
1975 1986	D.F.X.	2645	840* 700†	9500 12000	57.30mm Stroke Turbocharged version of the D.F.V. designed for Indianapolis type racing in the USA.
1978	K.A.A.	2400	240	7000	Rally version of the Opel "Ascona 400". In its first rally which was the Monte Carlo 1980, it came in 4th and 8th position. The next event, Swedish Rally 1980, it came first.
1979	O.A.A.	1600	170	7800	Formula Supervee. A Volkswagen based engine.

*80" Hg. boost
†48" Hg. boost

Year	Engine	C.C.	BHP	RPM	Remarks
1981	D.F.L.	3955	540	9250	90.00mm bore x 77.7mm stroke version of the D.F.V. designed for Group "C" World Endurance Championship for Makes. Manufactured under licence for Ford Motor Company. A 64.8mm. stroke version with a 3.3 litre capacity has been raced successfully since 1981 especially in Group 'C2'.
1982	D.F.V.ss	2993	500	11000	90.00mm bore x 58.8mm stroke version of the D.F.V. used in Formula 1.
1982	D.F.Y.	2993	500	11000	Similar to D.F.V.ss with a different head design which had a camshaft carrier integral with the cylinder head. Valve angles 10° inlet 12.5° exhaust. Five only examples were made for Tyrrell Racing. D.F.Y. 001 powered their Detroit GP win.
1983	B.D.R.	1601	120	6250	Based on the B.D.A. Sold as a conversion kit. 150bhp and 170bhp 1700cc versions are fitted in top of the range Caterham 7 sports cars.
	B.D.P.	1975	245	8500	Developed by Cosworth Engineering Incorporated for use in Midget racing. Fuel injected methanol burning, light alloy crankcase version of BD race engine.
1984	Y.B.B.	1995	204	6000	Turbocharged 4 valve per cylinder twin camshaft engine based on the European Ford Pinto cylinder block. Designed and developed specifically for the series production Ford Sierra RS Cosworth.
1986	Y.B.C.	1995	280	6000	Group A rally engine based on the Y.B.B.
1986	F3000	2993	420	8750	9000rpm rev limited F3000 spec. D.F.V.
1986	F1 Ford V6 (GBA)	1497			Formula 1 turbocharged Grand Prix engine developed for Ford.
1987	D.F.Z.	3495	560	10250	3½ litre normally aspirated Ford Formula One engine. (To compete for the Colin Chapman Cup in Grand Prix races).
1987	D.B.A.	3000	t.b.d.	t.b.d.	DOHC 16v version of Pontiac 'Iron Duke' engine for use in U.S.A. and to be sold in kit form.

FLOOR AREAS

Cosworth Engineering Ltd., St. James Mill Road, Northampton, NN5 5JJ		79,700 sq. ft.
Cosworth Engineering Ltd., Whittle Close, Park Farm Industrial Estate, Wellingborough, Northants.		29,000 sq. ft.
Cosworth Casting Ltd., 9 Hylton Road, Worcester, WR2 5SS	Foundry 1	12,500 sq. ft.
	Foundry 2	17,500 sq. ft.
	Ceramic Pumps	4,000 sq. ft.
Cosworth Engineering Incorporated, 23205 Early Avenue, Torrance, California, USA		7,080 sq. ft.
Brian Hart Ltd., Unit 5, Roydonbury Industrial Estate, The Pinnacles, Harlow, Essex CM19 5BZ		12,000 sq. ft.

DIRECTORS:

KEITH DUCKWORTH .... CHAIRMAN & CHIEF ENGINEER
RICHARD BULMAN .... MANAGING DIRECTOR
MIKE COSTIN .... DIRECTOR & DEPUTY CHIEF ENGINEER
BEN ROOD .... DIRECTOR & CHIEF INDUSTRIAL/PRODUCTION ENGINEER
HARVEY FOX .... DIRECTOR - (Wellingborough)
MIKE HALL .... DIRECTOR & CHIEF DESIGNER (COSWORTH ENGINES)
DICK SCAMMELL .... DIRECTOR & GENERAL MANAGER RACING ENGINES
BOB SMITH .... MANAGING DIRECTOR - COSWORTH CASTINGS LTD.
BOB GRAVES .... DIRECTOR - UEI/COSWORTH

(Picture credits)

Ford Motor Company
5 btm., 10, 17, 18 btm., 20, 22, 26, 27, 28, 30, 31, 35, 38.

William Kimberley
8, 9 btm., 11, 16 btm., 18 top, 19, 29, 33 top, 40 top 2, Back cover.

Geoffrey Goddard
4.

David Cundy
16 top.

Tracfoto
12/13